Soybean Storytelling

Translated by Diana Evans

Committee for the Establishment
of a Korean Soybean Museum

The original 'Soybean Storytelling' written in Korean
was translated into English by Diana Evans

Published April 5, 2017
Publisher Lee Cherl-Ho (Korea Food Security Research Foundation)
Sikanyeon, 109A Life Science Bld.(East), Korea University,
145 Anamro, Sungbukku, Seoul, 02842 Republic of Korea
Phone. 82-2-929-2751 / Fax. 82-2-927-5201
Email. foodsecurity@foodsecurity.or.kr
Printed in March 28, 2017 by Hanrimwon Co., Seoul.

국립중앙도서관 출판예정도서목록(CIP)
Soybean Storytelling / 한국콩박물관건립추진위원회 편. -- Seoul : Sikanyeon, 2017
Translated by Diana Evans
한영대역본
ISBN 979-11-86396-36-0 93400 : ₩20,000
콩[豆]
524.42-KDC6 / 633.34-DDC23 CIP2017007421

Soybean Storytelling

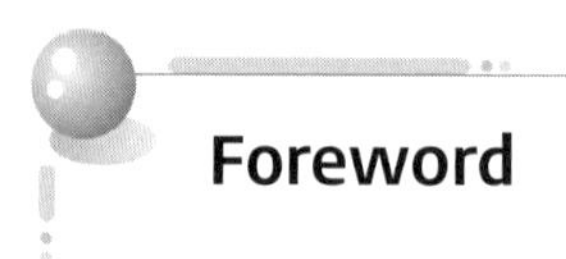

Foreword

The cultivation and usage of soybeans began in Korea, but because most ancient histories of Korea are intermingled with Chinese culture, this fact has not been properly disseminated to the world. Dr. Kwon Tai-wan, the founding director of the Korea Food Research Institute, formed the Committee for the Establishment of a Korean Soybean Museum in 2001 in order to begin the work on a museum that would inform visitors of the true history and use of soybeans. Below is the letter of intent penned at the Promotion Committee's inception:

Purpose of Promoting the Establishment of a Korean Soybean Museum

We wish to illuminate the research we have done, which indicates that in the history of humanity, Koreans were the first to plant soybeans and use them for food. Soybeans were used as food in Northeast Asia more than 3000 years ago; from the 3rd century BC soybeans were disseminated to the southern region of China, Southeast Asia, and Japan; they were introduced to Europe in the 18th century; and during World War II they were cultivated as a cash crop in the United States. Today soybean cultivation has spread all around the world.

It is known that soybeans supplemented with rice comprise a nutritionally complete meal, but recently it has come to light that soybeans also help prevent and cure cancer, heart disease, and other adult-onset diseases. Some cultures have already been supporting their health by consuming soybeans as a staple in their diets, and now in the West, where soybeans have long been used for oil or fodder, the cytological value of soybeans is being revealed, thus quickly ushering in an era of dietary soybean consumption. Due to the spread of cultivation and the variety of its uses, the soybean, thought it originated in Northeast Asia, has now become a global crop as well as a food item for people all around the world.

Soybeans, which helped sustain the health of our ancestors where they lived on the Korean Peninsula and in Manchuria, as well as that of their descendants through a long passage of time, have now been introduced on the world stage. The 21st century can be called the culture century, or an era in which culture is judged to be the conscience and capacity of nations. Many museums have been established here and there on this earth, each a symbol of its given culture, and yet nowhere is there a museum about soybeans. Is it not appropriate that a museum specializing in soybeans should be established in the origin nation of soybeans? By establishing a museum on an international scale that focuses on soybean culture and utilization techniques, we hope to broadly inform the global village of soybean culture and development, and by so doing contribute to humanity's health.

Going beyond simply amassing and exhibiting artifacts relating to soybean culture from days long past, research and analysis will be done on global data and literature relating to soybeans, thus creating a forward-looking, creative museum that provides informational exchange on an international scale and focuses on research education. There will be an all-weather greenhouse, a soybean products manufacturing lab, and even a soybean specialty café, such that everything about soybeans can be seen, learned, pondered, eaten, and experienced all in one place. In this way the museum will be self-supporting in its operations and development.

From this viewpoint we wish to establish a soybean museum that will revive our culture of the past thousand years and thereby bequeath the wise and creative spirit of our ancestors to future generations in the 2000s. We will marshal all of our abilities for the work at hand.

September 2001

Committee for the Establishment of a Korean Soybean Museum
Promoters: Kwon Tai-wan, Kim Seok-dong, Kim Seok-min, Kim Jun-young, Ryu Yong-hwan, Lee Cherl-ho, Lee Young-tack, Chang Hak-gil, Jeong Jang-seop, Jung Chai-won, Cho Se-young, Hong Eun-hi

With the support of research funds from the Daesan Agricultural Foundation in 1998, this work began with a study written by Kwon Tai-wan, Kwon Sin-han, Lee Cherl-ho, and Hong Eun- hi, "The feasibility study on the establishment of a World Soybean Center," which began an all-inclusive research project into the history and scientific development of soybeans.

In 2005 the committee, which included most of the major soybean specialists in the country, published the book *Soybeans* (15 chapters, 794 pages, published by Korea University Press). This reference work gathered all the information in Korea and abroad regarding the historical uses of soybeans, including ancient artifacts and remains, the early dispersion of wild soybeans, the history of cultivation, the nutritional profile of soybeans, soybean functionality, various foods and meals that use soybeans, and the present conditions and outlook of industrial uses and production distribution. The chapters with their authors are as follows:

Foreword, Kwon Tai-wan; 1) "History of the uses of soybeans," Lee Cherl-ho and Kwon Tai-wan; 2) "Prehistoric and ancient soybean remains," Cho Hyeon-jong; 3) "Sauce culture and earthenware," Shin Suk-jeong; 4) "The history of soybean cultivation," Hong Eun-hi; 5) "Soybean cultivars and breeding," Kim Seok-dong and Lee Young-ho; 6) "Characteristics of soybean processing," Kim Woo-jeong; 7) "History and current state of soymilk and soybean curd," Son Heon-su; 8) "Fermented soybean foods," Shin Dong-hwa and Lee Hyo-ji; 9) "The health functionality of fermented soybean products," Park Kun-young; 10) "Korean foods made with soybeans," Lee Hyo-ji; 11) "Soybean cuisine in other countries," Cho Jung-soon; 12) "The nutritive value and functionality of soybean foods," Seung Jeong-Ja; 13) "Industrial uses of soybeans," Chee Kew-man; 14) "Soybean oil and its byproducts," Lee Gyeong-il; 15) "The current state and future outlook of soybean production and distribution," Cho Se-yeong.

In 2008 the committee founded a website based on this book, the Soyworld Museum (www.soyworld.org). In 2011 it was decided to add a "Storytelling" section based on the book, and committee member Yoo Mi-kyung was tasked with this work. When the Soyworld Science Museum was established in the city of Yeongju, North Gyeongsang Province in 2014, "Soybean Storytelling" was vital to the exhibits and included in the museum catalogue. Committee members Lee Cherl-ho planning, Yoo Mi-kyung writing, and Kim Seok-dong, Moon Gap-soon, Lee Young-ho, Song Hi-seop, and Hwang Young-hyun supervised the bringing forth of the catalogue.

In 2015 the catalogue was translated into English by Diana Evans, who did graduate work in Korean Literature at Harvard University, and one year later the work of so many is now presented to the world. Supervision and correction of the press was made by Kim Seok-dong, Moon Gap-soon, Lee Young-ho, Song Hi-sup, Hwang Young-hyun and Hwang In-kyeong.

We express our great respect for and would like to sincerely thank Dr. Kwon Tai-wan and all those who devoted their time and efforts to collaborating on this work. We humbly give thanks for stellar historians in our midst.

April 2017

Lee Cherl-ho, 3rd Chairman,

Committee for the Establishment of a Korean Soybean Museum

Chairman, Korea Food Security Research Foundation

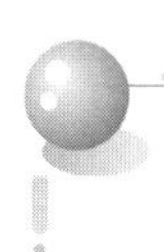

Translator's Note

Translating the catalogue for Soyworld Science Museum brought many delights of discovery. As I translated my way through the chapters I learned about the history, development, cultural uses, and functional health of soybeans, to name a just a few of the topics presented here. Before engaging this work I never thought soybeans could fill a museum, but I am pleased to confess I was wrong. In fact, I am thrilled there is a museum dedicated to soybeans, which play such a vital role in Korean culture and cuisine. Since the first day I set foot in Korea I have loved Korean food, and I particularly relished the tofu, especially during my time as a vegetarian. Korean cuisine continues to be my favorite in the world: the taste never grows dull or monotonous, thanks to the complex flavors of its sauces, all of which are based on soybeans. I learned much about the great variety of health benefits soy provides in all its dietary iterations, not least in fermented sauces. Translating the adages in chapter 2 that reference soybeans and sauces proved to be a challenge (even several Koreans I consulted could not shed light on the meaning of older sayings), but upon experiencing the flavors of Korean soy-based sauces, one can imagine the provenance of adages such as "A home with delicious sauces prospers."

My first loves are culture and literature, but I was pleasantly rewarded when translating chapter 5, which delves into the bioactive substances found in soybeans and soybean foods and how they interact with our bodies to our benefit. It was invigorating to find the appropriate translations for various isoflavones, amino acids, and other biochemical terms, even as I then researched those terms in my own language to familiarize myself with their function in the body.

As my deadline for submitting the translation approached, I shipped off chapters 4 and 6 to the Korean Translation Group in New York City and would like to thank them for their efforts and quick turnaround. I then edited these chapters to achieve a consistent tone with the rest of the work. I would also like to thank the helpful translators in the Korean forum at WordReference.com and friends who helped me with archaic or particularly tricky turns of phrase that could have been interpreted in different ways. Finally, I thank Chairman Lee Cherl-ho for inviting me to work on this project and for his patience while I labored methodically to present what I hope is a highly readable text in English that is also faithful to the original Korean.

January 2017

Diana Evans

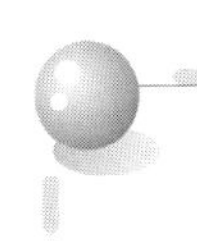

Contents

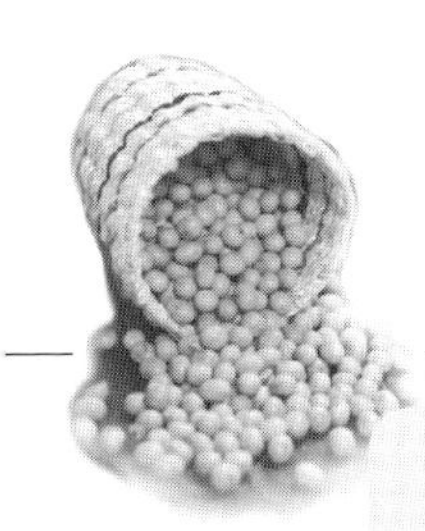

Chapter **01**

The Origin of Soybeans

1. When did we begin to eat soybeans?
2. Where did soybeans originate?
3. The image of soybeans over time
4. The story of soybeans through historical figures

1. When Did We Begin to Eat Soybeans?

It is fair to say that the history of humankind began with the effort to obtain food, and because the securing of food is necessary for human existence, this effort continues today. Indigenous food and culture developed according to the given climate of different regions in the world. According to archeological records, already in the Early Neolithic age (6000-3000 BC) in Korea, techniques for making yeast, brewing alcohol, and fermenting *kimchi* and *jeotgal* (salted seafood) had already reached a fairly high level of development. Globally, wild soybeans were found in Korea, the island of Taiwan, Japan, along the coast of the Yangtze River in China, Manchuria, and Siberia. Wild soybeans in Korea were evenly distributed across the country, from Jeju Island in the south to Hamgyeong Province, the most northerly province in North Korea. They grow wild in foothills, on grasslands, and in bushy or wooded areas. The remains of charred soybeans have been discovered at Paleolithic, Neolithic, Bronze Age, and Iron Age sites spanning the Korean Peninsula. There is a record of soybean pollen having been discovered in Yong cave at the Jecheon Jeommal site, which dates to the Late Paleolithic.[1)]

Soybeans in the Early Neolithic Age

Trypsin inhibitors in soybeans block the digestion of protein, which means ingesting raw soybeans will result in a severe case of diarrhea. That is why people of the Stone Age considered soybeans to be a poisonous plant. The Primitive Pottery culture developing in the area of the Korea Strait around 6000 BC gradually made soy into a foodstuff. The widespread traces of primitive pottery peoples reach to the north and south ends of the Korean peninsula and the Japanese archipelago in coastal areas. Today excavations along the coast of the Korea Strait are the foremost of their kind in the world and indicate that this area is the cradle of East Asia's pottery culture in Neolithic Age.[2)]

The native hunter-gatherers who lived along the seashore began to make pottery, which for the first time led to the development of a technique for storing and boiling water. This in turn led to the discovery that boiling soybeans made them edible, with none of the previous detrimental health effects. They placed sea water into their earthenware, collected marine plants or animals in the area, and boiled them up with vegetables, which resulted in *jjigae* (stew). Through this process they discovered a method of salt-making from sea water, which then led to preserving and fermenting foods with

1) Lee Cherl-ho, The food ways of Paleolithic men in the Northeast Asia and Korean Peninsula, Minjokmunhwayeongu (Korean Cultural Studies) 31:415-458(1998).

2) Barnes, G. L., The Rise of Civilization in East Asia: The Archaeology of China, Korea and Japan: Thames and Hudson Ltd., London (1993).

salt. Wet grains and roots stored in earthenware pots grew mold and yeast and turned into alcohol. Storing vegetables with sea water in earthenware jars caused the growth of lactobacilli, which created *kimchi* (a term that broadly encompasses fermented vegetables); doing the same with seafood created *jeotgal* (fermented shrimp, oysters or fish). It is thus that the Neolithic people along the coasts of the Korean Strait created the basis of a water-boiling culture and a fermentation culture.[1)]

Soybean and the Eastern Yi (Archers) Tribes

According to ancient literature from China, the Eastern Yi (or Dongyi, 東夷) tribes comprised the peoples living in Northeast Asia, including the Korean Peninsula, in Neolithic times. According to *Shuowen Jiezi* (one of the earliest Chinese-character dictionaries, written in 100 AD), the tribal name Yi(夷) comes from the combination of the characters for "big(大)", "arrow(弓)," and "man(人)" indicating tribes that use long arrows. Dr. Lee Sung-woo(1990) has concluded that the first tribe in human history to use soybeans as food was the Yemaek tribe, a tribe among the Eastern Yi (Archers) people at the period of the early state formation era of Northeast Asia (4000-2000 BC).[2)] Nomads in the north began to settle during the late Neolithic Age, particularly in the Baekdu Mountains of southern Manchuria and the northern Korean Peninsula, and there they began cultivating soybeans. Dr. Lee considers the ensuing Early

---

1) Lee Cherl-ho, The Primitive Pottery Age of Northeast Asia and its importance in Korean food history, Minjokmunhwayeongu (Korean Cultural Studies), 32:325-357(1999).

2) Lee Sung-woo, A study on the origin and interchange of Dujang (also known as soybean sauce) in ancient East Asia, J. Korean Society of Dietary Culture, 5(3):313-316(1990)

Yemaek Tribe

In Chinese documents from the Han dynasty (206 BC-220 AD), Joseon (ethnic Korean) tribes are referred to as Ye(濊), Maek(貊), or Han (韓, different from the Chinese Han, 漢). According to Choe Dong's 1963 *History of the ancient Joseon peoples*, the Eastern Yi tribes originated in the Central Asian kingdom of Babylonia about 3000 years ago, a group of who moved east through the southern region of Siberia, settled along the banks of the Songhua River, and eventually mixed with the aboriginal peoples there. This new offshoot moved south along the Liao River and into the Korean Peninsula, and was known as the Eastern Archers tribes. After Qin Shi Huang, Emperor of the Qin dynasty, built the Great Wall of China (225 BC), the Eastern Yi tribes in the region of China near the border and those in the Shandong Province region assimilated with the Han Chinese, and only those tribes living in southern Manchuria and the Korean Peninsula continued to be known as the Eastern Yi tribes.

Bronze Age (about 1500 BC) as the time when edible soybeans began to spread throughout the Korean Peninsula and Northeast Asia.

Early on in China the soybean was not called *daedu*(大豆) as it is today, but *yungsuk* ((戎菽) "warrior plant," or wild bean). This may be an indication that soybean does not originally come from the Chinese, but is believed to be a food of the Eastern Yi tribes from beyond the Great Wall-tribes the Chinese called Yung(戎) fighters, (warriors), or barbarians. On the other hand, the expansion of soybeans into the southern regions of China, Southeast Asia, and Japan occurred in about the third to fourth centuries BC. Charred remains of about 20 examples of rice grains and soybeans have been excavated from Neolithic lots in the village of Daecheon in Okcheon, North Chungcheong Province. Through radioactive carbon dating it has become clear that these date from the Late Neolithic, between about 3000-3500 BC, making them the oldest example of soybeans found on the Korean Peninsula. Meanwhile, other remains of soybeans that have come to light were found not in dwelling places, but rather in peat deposits, thus making it difficult to judge whether these were wild or domesticated soybean plants. Thus the history of the edible soybean has shared a pulse with the history of the Korean nation since its origin 5000 years ago.

2. Where Did Soybeans Originate?

When we think about the purchasing of goods, there is something known as "premium origin." Though two items may be similar in nature, a specialty place of manufacture can command a higher price and better sales. Setting aside questions of place of origin, the people of this land have processed and eaten soybeans for thousands of years. Dr. Gary Nabhan, the W.K. Kellogg Endowed Chair in Sustainable Food Systems at the University of Arizona Southwest Center, has said "We are the result of our ancestors' food and drink consumption," and "If our ancestors lived for a long time in one area, it is highly probable that we have become genetically adapted to the foods of that environment."[1] Viewed from that angle, the expression that soybeans and soy sauce flow through our DNA becomes more than simply a rhetorical device.

In order to discover soybeans' place of origin, we must investigate the results of researching literary sources about the cultivation of soybeans, the academic viewpoint of agriculture, archaeological evi-

1) Nabhan, G. P., Where our food comes from? Island Press, Washington D.C. (2008)

dence, and the current culture of soybeans. The opinion that soybeans originated on the Korean Peninsula and in Manchuria is founded on Japan's Dr. Fukuda's assertion that the division of wild soybeans led to many areas of origin.[1] The scholar most representative of the China origin theory is the United States' Dr. Hymowitz, and his assertions are based on the text in Shijing (The Book of Odes written in 11th-6th century B. C. in China). In this book, mention is made of a bean called "*suk*". In the 11th century BC China's northeastern region supported not just one tribe, but several Eastern Yi tribes. In the past the whole of Manchuria, China's northeast region and stage of the Korean people's activities, was known worldwide as the soybean's place of origin. However, Koreans now make their base on the Korean Peninsula and find that wild soybeans grow naturally there, and the cultivated genetic variation is so high that Korea is also considered one of the places of origin of the soybean.

Dr. Kwon Shin-han suggests that soybeans cultivated throughout the area of Manchuria in about 2500 BC spread to the Korean Peninsula and began to be produced as farm products in about 2000-1500 BC. He states further that "Korean native breeds of soybeans have seed sizes, seed coat colors, leaf shapes, harvest maturity times, and fat and protein contents that countries throughout the world currently report as characteristics their different soybean breeds possess. This attests to the fact that our ancestors' traditional cultivation system accumulated a great variety of soybeans."[2]

---

1) Fukuda, Y., Cytogenetical studies on the wild and cultivated Manchurian Soybeans (Glycine L.), Japanese J. Botany (1933)

2) Kwon Shin-han, The origin of soybeans, Korea Soybean Digest, 2:4-8(1985)

Investigating the Literature

A passage in *The Lost Book of Zhou*, which was published in about the 6th century BC, reads "The nomadic warriors (*san**yung***) are the Northeastern Yi tribes. The warrior beans (***yung**suk*) that grow there are large beans." Sima Qian's *Shiji* (*Records of the Grand Historian*) states, "As soon as the state of Qi heard that nomadic warriors attacked the state of Yan in 623 BC, Duke Huan of Qi saved Yan by conquering the enemy and chasing them north all the way to Guzhu; from thence he brought back warrior beans. Duke Huan presented this warrior plant to the neighboring state of Lu." *Guanzi* reads, "Duke Huan of the state of Qi pushed the warring nomads north and brought back winter onions and warrior beans, which now cover the land." Looking at the three historical reference works above, the "beans of the warrior nomads" seem to be the earliest appearance of soybeans in recorded history. According to Pusan University Professor Choi Duk-kyung, "If we reference China's Qi-based historical documents, the soybean began to be cultivated at the beginning of the Zhou dynasty in the northeast region of China; after the time of the Spring and Autumn Period (771-476 BC) the plant was disseminated throughout the northern part of China; and finally, after the Qin and Han, cultivation expanded to all of China, and the plant name changed to that of "soybean(大豆)."[1]

1) Choi Duk-kyung, The origin theory of soybean cultivation and the Korean Peninsula, Jungguksayeongu, vol. 31, Jungguksayeonguhoi, Seoul (2004)

Earliest appearance

According to cultural anthropologist Jared Diamond in his book *Guns, Germs, and Steel*, whether Europe's oldest human remains or Mexico's first charred corn crop remnants, "the earliest appearance" of a given item always exists. In *The Lost Book of Zhou*, *Records of the Grand Historian*, and *Guanzi*, the references to nomadic warriors indicate that the banks of the Daling River in Chaoyang, Manchuria is where soybeans were originally cultivated, and so this becomes their earliest appearance. In other words, the current area of Chaoyang was where the nomadic warriors roamed. In *Memorabilia of the Three Kingdoms* (*Samguk yusa*) the original capital city of Gojoseon is Asadal, which when translated into Chinese characters becomes Chaoyang. The area of Chaoyang encompasses excavations of Liaoning bronze daggers, representative relics of Gojoseon.

The History and Spread of Soybeans

It is generally understood that through 700 AD soybeans spread across the whole region of southern China and Southeast Asia. It is judged that the spread of soybeans to Southeast Asia is interrelated with the history of the southward migration of overseas Chinese. Already from the 4th century AD China's poor had become known as "the nation that came south" in Southeast Asia, but from the 9th to 13th centuries it is supposed that many Chinese effecting a large-scale southward migration resulted in establishing soybean cultivation and methods of soybean use throughout the Southeast Asia region, including the Indochinese Peninsula. Soybeans were introduced to Europe in 1712 when German scholar Engelbert Kaempfer returned from Japan and brought back soybeans for his personal interest. Official introduction occurred in 1739 when French missionaries returned from China with soybean seeds, which were for the first time planted in the Paris Botanical Gardens. Once soybeans were planted in the British Botanical Gardens in 1790, they were tested for cultivation in the East and West African British colonies.

There are two paths by which soybeans came to the United States. The first was in 1764, when a sailor with the East India Company named Samuel Bowen returned from living in Canton. The second is when in 1770 then-ambassador to France Benjamin Franklin sent a letter from England containing soybean seeds to his home in Philadelphia. Although soybeans were introduced from the Orient to the West in the late 18th century, another century would pass before they garnered interest as an economic product. After the first Opium War (1840-1842) American agriculturalists observed how the Chinese used soybeans for food and began a full investigation into the cultivation techniques of this "cash cow of the field." As the 20th century ushered in the first and second world wars, soybeans began to be used as fodder and green manure crops, so much so that they were known as the "Cinderella crop" and the "miracle crop," becoming the world's most important staple crop.[1)]

Currently the number one producer of soybeans, in the early 20th century the United States gathered many soybean genetic resources from China, Japan, and Korea. From 1929-1931 two US Department of Agriculture workers, Palmon Dosett and William Morse, were dispatched to Asia to retrieve soybean genetic resources. From Korea they gathered 3,379 specimens (74%), from China's

---

1) Lee Cherl-ho and Kwon Tai-wan, History of the use of soybeans, *Kong* (Soybeans), Committee for the Establishment of a Korean Soybean Museum, Korea University Press, Seoul, 3-44(2005)

northeast region 622 specimens (14%), and from Japan 577 specimens (13%). These numbers are indicative of the vast diversity of soybean genetic materials found in Korea.[1)]

The Dissemination of Cultivated Soybeans

- 25th century BC, Manchuria region
- 20th-15th centuries BC, Korean Peninsula
- 4th century BC, southern China
- 3rd century BC, Japan
- 9th century AD, Southeast Asia
- 1712, record of Kaempfer introducing soybeans to Germany
- 1739, experimental cultivation at Paris Botanical Gardens
- 1765, Samuel Bowen plants soybeans in Savannah, Georgia
- 1770, Benjamin Franklin sends soybeans to Bartram's Garden in Philadelphia
- 1786, experimental cultivation at German Botanical Gardens
- 1790, experimental cultivation at London Botanical Gardens
- 1852, Commodore Perry brings back soybean seeds from Japan
- 1898, under USDA supervision introduction of 65 varieties from 7 countries
- 1929-1931, William Morse collects seeds on a "plant expedition to the Orient"
- 1935, soybean output of the US reaches 50% of world production
- Late 20th century, Brazil, Argentina, and other Latin American countries cultivate soybeans
- 2013, Afghanistan begins cultivating soybeans

Academic Crop Science Viewpoint

In terms of botany, an important index used to estimate the birthplace of a crop is the extent to which the wild species is distributed. In academia, place of origin is defined as a single place in which wild soybeans, intermediary forms, and cultivated soybeans all grow, and

1) Kim Seok-dong and Lee Yeong-ho, Soybean varieties and breeding, *Kong* (Soybeans), Korea University Press, Seoul, 137-216(2005)

those places which correspond to this definition are Manchuria and the Korean Peninsula. However, one reason the theory of Korea as place of origin has gone unnoticed by scholars is that research materials on Korean soybeans are scarce. For example, an American scholar might study Chinese or Japanese soybean genetics and so produce articles about them, but Korean soybeans are omitted or barely investigated (usually due to lack of familiarity with Korean), which results in little to no information produced about them.

Recently Seoul National University (SNU) Professor Lee Yeong-ho presented his findings after having ordered various genetic varieties of soybeans from the southern region of the Korean Peninsula, which have been excavated in great numbers as Bronze Age charred soybeans and other soybean remains. He said, "After the earth's greatest glacier growth (11,000 BC), the Chinese continent separated from the Korean Peninsula, and the people on the peninsula independently developed the cultivation of pulse crops according to their own selection and clime. SNU Professor Lee Seok-ha has also said, "In the future more pulse remains will be discovered, and once a molecular clock that will analyze DNA to establish the evolution of soybeans is developed, it will become widely known that the Korean Peninsula is the place of origin of the cultivated soybean."[1)]

1) Kim, M. Y., et al., Whole-genome Sequencing and intensive analysis of the undomesicated Soybean (Glycine soja Sieb, and Zucc.) genome, PNAS, 107(51):22032-22037(2010)

The Archeological Community on Excavated Relics

Wonkwang University Professor An Seung-mo has said, "From the Early Stone Age wild pulses were used, but pulses that have been positively identified as cultivated come from excavation remains dating from the 1,000s BC. Archeologically speaking, soybeans are assumed to have originated in the area of Manchuria and on the Korean Peninsula. Soybean remains have been excavated at Bronze Age sites on the Korean Peninsula and in the region of Jilin (China) from at least 3,000 years ago, while discoveries in China date mostly from the Han dynasty (220 BC-220 AD), and those in Japan from the Yayoi era after about the 4th century BC."[1)]

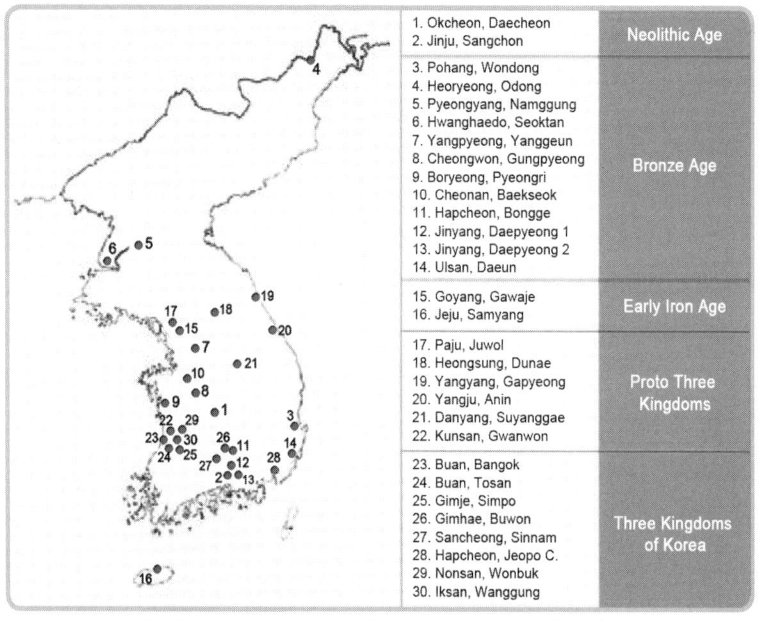

Soybean excavation sites in Korean Peninsula

1) Cho, H.-J., Soybeans in Prehistoric and ancient remains, *Kong* (Soybeans), Korea University Press, Seoul, 45-80(2005)

Excavated Remains of Soybeans on the Korean Peninsula

Numerous remains of soybeans are currently being discovered in dozens of excavation sites across the Korean Peninsula, which range in time from the Neolithic Age to the Bronze Age and Early Iron Age, as well as the Proto Three Kingdoms and the Three Kingdoms Periods.[1)]

Neolithic Age (before 1500 BC)

Remains of soybeans and other related foodstuffs of the Neolithic Age can be found at the Daecheon Village, Okcheon site and the Sangchon Village, Jinju site. However, the charred rice, barley, wheat, foxtail millet, and other grains reported found with the soybeans are not mentioned in the official report of the Daecheon Village site, whereas the Sangchon Village report references the wheat, barley, foxtail millet, broomcorn millet, and other grain materials along with the remains of the pulse family. These remains have become known as the earliest soybean materials found in Korea.[2)]

Bronze Age (1500 BC-300 BC)

By the Bronze Age rice farming had taken root on the Korean Peninsula, and food culture traditions such as that of raising the "Five Sacred Grains" (or "crops") had become established. Bronze Age soybean remains have been found on 12 sites around Korea, including Odong, Heoryeong, North Hamgyeong Province; Nam-gyeong, Pyeongyang; Seoktalli, Hwanghae Province; Yanggeun Village, Yangpyeong, Gyeonggi Province; Gungpyeong Village,

---

1) An, S.-M., Archeological records on the origin of legumes cultivation in East Asia, Korea Soybean Digest, 19(2):24-33(2002)

2) Lee Yeong-ho, Park Tae-shik, Origin of legumes cultivation in Korean Peninsula by View Point of excavated grain remains and genetic diversity of legumes, Nongupsayeongu (Agricultural History Studies), 5(1):1-31(2006)

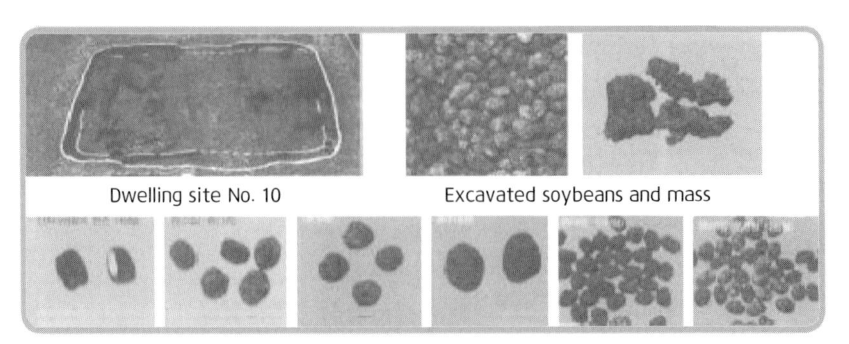

Bronze Age dwelling site No.10, Wondong, Pohang

Cheongwon, North Chungcheong Province; Pyeongna Village, Boryeong and Baekseok districts, Cheonan, South Chungcheong Province; Bonggye Village, Hapcheon and Daepyeong, Jinyang of South Gyeongsang Province; Daun district, Ulsan.

Early Iron Age (300 BC-1 AD)

The dissemination of iron farm tools precipitated a rapid increase in agricultural productivity. The soybean resources of this era are found in remains excavated in Gawaji Village, Goyang, Kimpo and Samyang districts, Jeju. The remains discovered at the Gawaji site include rice, gourds, and peach tree seeds, while charred remains of rice, wheat, barley, and more were found at the Samyang site.

Proto Three Kingdoms Period (1 AD-300 AD)

Six sites from this era have confirmed remains of soybeans: Juwol Village, Paju; Dunnae, Hoengseong; Gapyeong Village, Yangyang; Anin Village, Myeongju; Suyanggae, Danyang; and Gwanwon Village, Gunsan. Types of beans found at these settlement sites include soybeans, adzuki beans, and mung beans. Aside from the black-eyed peas, the soybeans, adzuki beans, and mung beans excavated at each site indicate that since the Bronze Age legumes had become a set part of the diet and had been cultivated continuously.

Three Kingdoms Period (300 AD-668 AD)

Soybeans have been detected at eight sites from this period, including Bangok Village and Tosan Village, Buan; Simpo Village, Gimje; Wonbuk Village, Nonsan; Wanggung Village, Iksan; Buwon district, Kimhae; Sonam Village, Sancheong; and Jeopo Village, Hapcheon. Among the grains excavated at the Wonbuk Village site, the presence of *saepat*, a wild, smaller ancestor of the adzuki bean, indicates that even while cultivated beans were being consumed, wild varieties of legumes were still used as food.

3. The Image of Soybeans Over Time

It is not well-known how people lived in Gojoseon, nor how rich or powerful it was. We can only guess at their cultural level by examining Bronze Age artifacts like dolmens, Liaoning bronze daggers, and comb-patterned earthenware. Gojoseon was followed by Buyeo and Goguryeo. Whatever the nomenclature of Korean people's culture–Gojoseon, Goguryeo, Balhae, Unified Silla, Goryeo, Joseon–soybeans have for thousands of years shared in the fate of the people as their "national food."

Gojoseon

Beans (du, 豆) in Oracle Bone Script

In the cultural sphere of Chinese characters, that which represents soybeans is widely used and written as 豆 (*du*). The first usage of this character is found at a Shang (China) excavation site, written on an oracle bone. It is known that the Eastern Yi tribes were the driving force behind establishing the Shang dynasty (1600 BC-1046 BC). During the Shang dynasty Gojoseon (2333 BC-108 BC) existed in China's northeast region. The "*du*" first appearing on oracle bones did not refer to soybeans, but indicated a bowl used for ancestral offerings. So when did the *du* that meant 'a footed dish used in ancestral rites' turn into the *du* that now means soybeans? Ancient peoples living 3,000-4,000 years ago witnessed the close of hunter-gatherer society and began a settled lifestyle. Food obtained from the ground at harvest time would naturally be offered in thanks to the gods, much like the prayer of thanks offered before the first harvest in the American New World that became the origin of the Thanksgiving holiday. Whether we perform a ceremony or not, and whether we know it or not, soybeans overlap with rites, offerings, gods, ancestors, and giving thanks. The importance and weight of soybeans is never light.

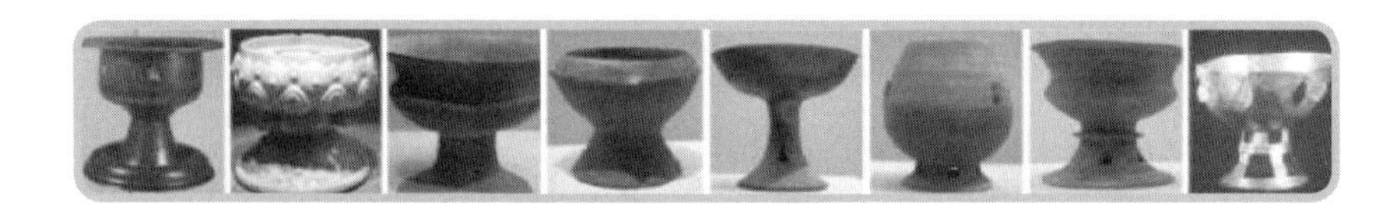

Oracle Bone Writings

During China's Shang dynasty, writing was done by engraving on the inside of a turtle shell or on the shoulder bone of a cow. In 1899 a Beijing bureaucrat by the name of Wang Yirong, while taking medicine to control his malaria, discovered a piece of turtle shell mixed in with his drugs. On the inside he found engraved markings resembling the shape of Chinese characters; this turned out to be a 3,000-year-old turtle shell excavated from what soon became a Shang archeological site.

Yungsuk

In terms of Northeast Asian literature, soybeans appear for the first time in *Shijing* (the earliest book of Chinese poetry) as *suk*(菽). In this text, *du*(豆) refers to a dish used in ritual offerings, and *suk* means soybeans. However, during the Han dynasty (206 BC-220 AD) it was said that the pod of the soybean (*suk*) resembled the wooden dish (*du*) used for ritual offerings, and thus *suk* eventually became *du*. In history books, the first instance of soybeans appeared in the 7th century BC in Guan Zhong's *Guanzi*, in which soldiers from the state of Qi brought back "wild" soybeans (*Yungsuk*). Yet the grammatical construction of the soldiers bringing back the soybeans indicates that these were not wild soybeans they found growing randomly in a field. Looking at history texts from 3,000 years ago that mention soybeans (*Yungsuk*), we can surmise that when Qi invaded the Eastern Yi tribes, they had already begun to domesticate the wild soybean. Perhaps it was even because of the soybeans that war broke out.

Goguryeo Dynasty

Goguryeo's Fermentation

Goguryeo's fermentation techniques were already known in neighboring lands, and the people there called the characteristics of the fermented product "Goryeo smell" (*Goryeo chwi*). Towards the end of the 3rd century, Zhang Hua, a Jin dynasty official and poet, wrote in his *Record of Diverse Matters* (*Bowuzhi*), "There is a foreign method of making fermented soybean cake, or in other words, in foreign lands there is fermented bean cake." The phrase "Goguryeo fermentation" (*Goguryeo seonjangyang*) appearing in the third-century text *Records of the Three Kingdoms, Book of Wei, Biography of the Eastern Yi Tribes* (*San guo zhi, Wei shu, Dongyi zhuan*) indicates that "The people of Goguryeo excel at making fermented foods." In China's first agricultural techniques texts, *Quimin yaoshu*, written in 532, four types of soybeans are mentioned: Goguryeo yellow beans, Goguryeo black beans, Swallowtail beans, and peas. Swallowtail beans and peas are not actually soybeans. In the sixth century there was no other soybean than the original Goguryeo bean.

Soy Sauce from the Tomb of Deok Heung-ni

"Tomb construction required the labor of 10,000 people, and every day cows and sheep were slaughtered, and the alcohol, meat, and rice could not be consumed all. In order to never run out of soy sauce, a storehouse of it was continuously maintained for the breakfasts of visitors to the tomb."[1] Deok Heung-ni's tomb, which dates from the Goguryeo dynasty, is located in Nampo City, Gangseo Precinct, Deok Heung district and is designated as North Korean National Treasure Cultural Relic #156. This ancient tomb was built in 408 (17th year of Gwanggaeto the Great) and belongs to a grand vizier of Goguryeo named Jin. Various customs are inscribed in over 600 characters on the walls and ceiling of the tomb, and once this writing was completed on the second day of the second month in the year 409, the door of the tomb was closed. Currently there is a dispute over whether Jin was Chinese or a Goguryeo native, but regardless this tomb is a precious relic from which the culture and customs of early 5th century Goguryeo can be discovered. The owner of the tomb, how-

1) Anak No. 3, Deok Heung-ni old tumb.

ever, left a storehouse full of soy sauce and disappeared from history. At this time King Gwanggaeto the Great had come to power, and the Goguryeo dynasty was enjoying its heyday.

Unified Silla

Ceremonial Wedding Food in the Era of King Sinmun

In the Silla annals found in *Samguk Sagi* (*Memorabilia of the Three Kingdoms*), King Sinmun married Kim Heum-un's daughter in the third year of his reign (683), and there is a list of ceremonial food items that were sent to the bride family, including rice(米), alcohol(酒), oil(油), honey(蜜), soy sauce(醬), fermented soybean cakes (豉), dried meat(脯), etc., for a total of 135 items. Through this we can see that fermented soybean cakes and sauces were important enough to be among the list of foodstuffs sent to the family of the king after a wedding. The fermented soybean cakes in the annals appears to be today's *meju*(soybean fermentation starter) It is said that "the history of *meju* is the history of jang(醬)", and these cakes hold the prominent position of originating fermentation in the making of Korea's traditional soybean sauces.

Samguk Yusa: Heung Nyun Buddhist Temple's Fermented Sauce

In *Samguk Yusa* (57 BC-935 AD) a touching folk tale by Kim Hyeon-gamho tells of a tiger that transforms into a maiden and marries Kim Hyeon, only to end by choosing death for his sake. Resnlting from the story, it has been said that "people who are bitten by a tiger apply Heung Nyun Temple's fermented soybean sauce to the wound." Also, not long ago Koreans would spread fermented soybean paste on the affected area after bee stings or other similar wounds. According to this text, evidently this familiar folk remedy was employed much longer ago as well, using the easily obtainable fermented bean paste made by a local temple. After King Beop Heung officially recognized Buddhism in 528, the first Silla Buddhist temple was established under King Jin Heung in 544. King Jin Heung designated this temple "The Great King Heung Nyun Temple" and later became its chief monk. Much later, during the Joseon dynasty, the fact that fermented soybean *meju* were used to make aged soy sauce at the palace, and tofu for the royal tombs depended on temple soybeans, shows that Buddhist temples and soybeans, and again the royal family and soybeans, formed inseparable relationships from very long ago.

Balhae Kingdom

Chaekseong Fermented Soybean Cakes

One of Balhae's well-known products was called "Chaekseong *si*" (*si*, 豉). There are differing opinions about the meaning of *si*, whether it referred to an early-type of fermented soybean product before they are divided into soy sauce and soybean paste or whether it referred instead to the dried *meju* or fermention starter before the steeping process to make soy sauce and *doenjang* (fermented soybean paste) (the latter opinion is more prevalent). Chaekseong covers the area of the Pallyeon Fortress site in present-day Hunchun City, Jilin Province (in northeast China, bordering North Korea). Both in olden days and today, the area is famous for its soybeans. During the Kingdom of Balhae (698-926), the type of produce farmed depended on the climate, soil, and topography; in warm climates millet, soybeans, and rice were cultivated, and in colder climates millet and soybeans were grown. In *New Tang History* (618-907) we read "A product we hold dear is Chaekseong *si*."

Is *si* (豉) soy?

In *Liji* (*Book of Rites,* 450 BC-100 AD) and *Chu Ci* (*Elegies of Chu* 300 BC) both, the question arises: "What is this thing called *si*?" In Chinese documents written before the Qin dynasty (秦代, 221-209 BC) there are no mentions of *si*, but by the Western Han dynasty (206 BC-24 AD) *si* has become an important product. In the *Shiji*, or *Records of the Grand Historian* (90 BC), mention is made of a thousand yeasty soups and salty *si*. Chin dynasty(晋代) author Changhua's *Natural History* presents *si* as a product of foreign origin: "One can find *si* in foreign lands," as does Lee Si-jin (1518-1593) in his text *List of Botanicals*. The Song dynasty (960-1126) text *Xue qi zhan bi*, regarding a scholar's teachings, reads, "There is no mention of *si* in *Buddha's Nine Discourses*, but it is used in the local vernacular." In Balhae, which succeeded Goguryeo, *si* was famous, and at the beginning of Unified Silla, too, *si* was known. However, in the Joseon dynasty the term meju (cakes of dried, fermented soybean) gained currency, and as words like *cheonggukjang* (fast-fermented soybean paste) and sometimes *jeongukjang* (an old variation in spelling) grew in frequency, *si* began to disappear. If read in contemporary Chinese, *si* sounds like *chǐ* or perhaps *shì*, but in old Chinese it would be pronounced (by Koreans) *seui*. In a Middle Korean dictionary too, *si* is pronounced *seui*. In English the word is **soy**bean, in Japan a fermented sauce made of soy is called **sho**yu. This coincidence of "soy" sounds seems rather unusual.

Goryeo Dynasty

Famine Relief Foods

In *Sikhwaji* (History of the economy) in *Goryeosa* (*The History of Goryeo*, 1058) it is recorded that in 1018 (9th year of Hyeonjong) after the Khitan invasion, rice, salt, and sauces were provided for the people who were starving and suffering from cold. In 1052 (6th year of Munjong) there is a record of rice, millet, and *doenjang* (fermented soybean paste) given to the approximately 30,000 people left starving after the monks' uprising against the military government. Famine relief foods to help the starving people are those foods necessary to sustain life. The fact that soy sauce and *doenjang* were included in the list of basic foods to relieve famine, along with rice, millet, and other grains, indicates that already by the Goryeo dynasty the idea that fermented sauces were necessary food products had taken root.

Joseon Dynasty

The Heyday of Fermented Sauce

Joseon's foremost cookbook, *Gyuhapchongseo*, loosely translated as Women's encyclopedia, was written in 1809. The book records how to make fermented sauces, how to select an auspicious day for making them, which items one should not include, storage methods, and medicinal effects. Soy sauce is effective for helping prevent edema and maintaining one's energy levels. In *Guhwangchwalyo* (Compendium for famine relief) and *Guhwangjeolyo* (Famine relief urgency) detailed instructions are provided for using soybean leaves in the making of soy sauce. Joseon cookbooks such as *Yorok* (1680), *Jubangmun*, and *Siuijeonseo* (late 1800s) all include recipes for making soy sauce. *Sallim gyeongje* (Farm management, end of 17th century) and *Jeungbo sallim gyeongje* (Farm management supplement) include 25 and 45 articles, respectively, on making fermented sauces and pastes. Phrases like "Sauce is boss" and

"The head of the family understands the importance of making fermented sauce and knows it must be left idle for a long time in order to have a fine sauce or paste" are found in these texts, indicating that fermented sauce is the basis of all flavors.

Fermented sauce recipes in *Jeungbo sallim gyeongje* (Farm management supplement)

Various fermented sauces, fermentation methods, mixing various ingredients when fermenting soy sauce, things to avoid when making soy sauce, properly packing a fermenting crock, clear ferments, reviving a sauce that has lost its flavor, fish and meat ferments, fermented soybeans with soybean *meju*, sauce made from a *meju* of fried soybeans and flour, fermented buckwheat sauce, fermented barley sauce, elm fruit sauce, red bean paste sauce, green bean sauce, quick fermented sauce, quick clear fermented sauce, red pepper paste, quick red pepper paste, *meju* of eggplant and mixed vegetables, summer eggplant and mixed vegetable sauce, *jeonsi* sauce (commonly known as *jeongukjang*, or "wartime sauce"), *cheongtae* and *jeonsi* sauce, *susi* paste, egg sauce, fried paste, boiled-down sauces, rice cake paste, freshwater sauce, thousand-mile sauce, etc.

Early Modern Times

Directive to Ban Rice Exports

When food is discussed in the framework of a crisis, one may hear about the "weaponization" of food. In early modern times, in the period known as the enlightenment era, an incident banning rice exports to Japan shows how food and war can become intertwined. In 1889 the governor of Hamgyeong Province, Jo Byeong-sik, declared a ban on exports or any sale of rice or soybeans to the Japanese. Japan demanded retraction of the ban and reparations, and the ensuing farmers' complaint of lack of food led to the Donghak Rebellion. The Daewongun, or prince regent at the time, requested help from China, while Queen Min asked for help from the Japanese army. Thus began a war between China and Japan on Korean soil, which ended in victory for Japan, leading to the exploitation of Korean rice, soybeans, and other grains. For 12 years, from 1917-1928, the exportation of grains to Japan came to fully half of Korea's entire output. In 1931 Japan invaded Manchuria and ruled there for the next three years. After bombing Pearl Harbor in 1941 and thus conflating World War II, we can guess whence they secured their military provisions.

The Great War and Soybeans

At the turn of the twentieth century, soybeans spread around the world and became an important grain during the World Wars. During World War I the Western world experienced a severe shortage of food. In 1918 England implemented food rations and began making bread by mixing soybean flour with potato flour. In 1918 in the United States, too, a movement began to encourage people to save flour, meat, and fat by using soy flour. Today the US is the number one producer of soybeans, and it all started with "In order to win the war, plant soybeans!" and the ensuing rapid expansion of soybean production lands. In Korea the self-sufficiency rate of soybean production stands at only 10%. In times of emergency, barley or millet can replace rice, and soybeans can replace meat. However, there is no crop or food product that could substitute for soybeans, so it seems that soybeans can be an important weapon.

4. The Story of Soybeans through Historical Figures

In this section we have chosen 33 famous persons in the East and West that have a connection with soybeans. Some of the characters made no direct contribution to soy, but they overlap with important events in the history of soybeans, or the background of the period in which they lived is interconnected with soybean. Therefore, we posit that the figures listed here contributed to the spread and expansion of soybeans around the world. At times the exact period is clearly noted, but when the occurrence of a certain event is not known for certain, the figures are listed in chronological order of their birth.

Duke Huan of Qi (r. 685-643 BC): The Propagation of Soybeans in China

The first mention of soybeans in a historical text occurred during the time of Qi's Duke Huan. Duke Huan was the foremost of the famous Five Hegemons in that era. At his side was the best bureaucrat in China, Guan Zhong, and he helped Duke Huan rise to the position of Prime Minister. In *Guanzi*, which Guan Zhong and his disciples wrote, there is a part that reads, "After his campaign against the nomad warriors (*sanyung*), Duke Huan of Qi brought back wild beans (*yungsuk*)." As explained above, *yungsuk* referred to the soybeans used by the nomadic warriors. At the request of the state of Yan, the Duke of Huan attacked the nomad warriors that were harassing Yan, and on his way home he picked up some beans and disseminated them in China.

Confucius (551-479 BC): Sings of Soybeans in The Book of Odes

Upon careful inspection of the charred remains of beans at 3,000 to 4,000-year-old excavation sites, it appears possible that the cultivation of soybeans significantly predates the state of Qi. Instances of the words *suk* and *du*, both referring to soybeans, occur 2,500 years ago in *The Book of Odes*, compiled by Confucius. At that time soybeans were called *suk*, and *du* referred to *jegi*, or the dishes used for offerings in ancestral rites. King Mu of the Zhou dynasty intones in his poem "The Public," "You planted soybeans, and the soybeans grew, swaying in the wind," and "Offerings are placed on ritual dishes, the saucer and the large plate, too." After a long period of time, *jegi* (the offering dishes) became synonymous with *jemul* (the food offerings placed upon the dishes), and the two became *jemul*, which is the old word for *du*, or soybeans.

Liu An (178-122 BC): Soybean curd's Progenitor

Liu An, the grandson of Han dynasty founder Liu Bang, was raised to become Prince Huainan. In Korea the first mention of soybean curd occurred in the late Tang, early Song period literary work *Cheongirok*. It appears that soybean curd entered China during the late Tang dynasty (618-907) due to exchanges with nomads in the northern regions. Yet Li Shizhen's Ming dynasty book *Compendium of* Materia Medica, published in 1596, states, "Liu An invented soybean curd," causing Liu An to become well-known as the 'progenitor of soybean curd'. Of course, doubts may arise when as many as 1,500 years come between the time when someone is supposed to have invented soybean curd and the time when that event is recorded in a literary work.

Cao Zhi (192-232): "Boiling the Beans, Burning the Stalks"

"Boiling the beans, burning the stalks" literally refers to boiling soybeans in a cauldron while the stalks provide the fire underneath the pot, and metaphorically to infighting among brothers. In the late 14th-century *Romance of the Three Kingdoms*, a fictional re-imagining of the classic text *Records of the Three Kingdoms*,

Cao Cao, the last victor of the three kingdoms period, has two particularly notable sons, Cao Pi and Cao Zhi. Cao Cao dotes on his younger son, Cao Zhi, because of his outstanding literary talent. He wants Cao Pi to always look after and protect his younger brother. When Cao Pi ascends to the throne, he worries about his brother's political ambitions and demands a literary test of his brother's skills, on pain of death. Cao Zhi must compose a poem during the seven footfalls it takes to approach the throne. Cao Zhi succeeds and saves his life with the following:

> They were boiling beans on a beanstalk fire;
> Came a plaintive voice from the pot,
> "O why, since we sprang from the selfsame root,
> Should you kill me with anger hot?"
> (Translated by C. H. Brewitt-Taylor)

Bae Hyeong-gyeong (874-936): Bean Sprout Story

Bean sprouts have been consumed in Korea for a very long time, but the first formal mention of them in writing began in the early Goryeo dynasty. In 935 as Goryeo's King Taejo established a new kingdom, there is a story that during a famine, General Bae Hyeon-gyeong helped alleviate the soldiers' suffering by immersing soybeans into a stream and allowing the soldiers to eat the resulting sprouts until their bellies were filled. The first mention of soybean sprouts in a text is found during the time of King Gojong (1214-1260) of Goryeo in *First Aid Prescriptions Using Native Ingredients*, or *Hyangyak gugeupbang*. In an entry called "Yellow soybeans," soybeans are barely sprouted and then dried in the sun for use as medicine.

Yi Saek (1328-1396): Dubu in the *Collected Works of Mogeun*

The first time soybean curd (*dubu*) appeared in the literature was during late Goryeo. It is said that *dubu* was passed down from the Mongols during this time of great exchange with the Yuan dynasty. Although we do not know exactly what route soybean curd took to Korea, its first mention in a text appears in Yi Saek's *Collected Works of Mogeun* (Mogeun is Yi's penname, meaning Peaceful Recluse). Early on in his life, Yi Saek studied abroad in the Mongol Yuan dynasty and served in various public offices there. He appears to have become very fond of *dobu*, as he mentions it in several different poems and writings in his collection. Phrases concerning soybean curd include, "We added taro to our *dubu* side dish," and "Fry the *dobu* in oil and then chop it up and place it in the soup to boil." Also, "Ah, *dobu* tastes like freshly sliced fat," and "If your teeth are loose, *dobu* is just the thing!"

Yi Seong-gye (1335-1408): The Legend of Gochujang

Gochu, or spicy red peppers, are said to have been introduced to Korea in the 16th century after the Imjin War. Interestingly, though, there is a legend about the founder of the Joseon dynasty, Yi Seong-gye, in conjunction with "Sunchang *gochujang*" (red pepper fermented paste most famously produced in Sunchang, North Jeolla Province, South Korea). The question is whether Yi Seong-gye actually ate the same kind of *gochujang* we eat today. After ascending to the throne, Yi Seong-gye sought the location of Manil Temple in Sunchang County, where he had lived with his mentor, Muhak, and once stopped by a farm to help make *gochujang*, after which he had lunched in the farmer's home. Yi could not forget the taste of that delicious meal and set out to discover the source. Thus was born the legend of *gochujang* and the founder of the Joseon dynasty. In a 1740 cookbook called *Somun saseol*, abalone, king shrimp, mussels, and more are prepared using the "Sunchang *gochojang*" [sic] method, thus indicating something special about Sunchang *gochujang*.

King Sejong (r. 1397-1450): Dubu-making Women Sent to the Ming Emperor

Soybean curd is said to come to Korea via the Yuan dynasty in the 12th century. Ironically, in the 15th century "*dubu*-making women" from Korea were popular among the Chinese imperial family. In the 10th year of King Sejong (1428), the bureaucrat who went to China as an envoy with gifts of *dubu*, among other things, received a government position there. Also, in the 16th year of Sejong (1434) an envoy brought back a royal letter from the Chinese emperor that reads, "The women you sent previously who make side dishes and other food have brought harmony to our meals...their Soybean curd becomes more and more exquisite.... Please select 10 more such women and send them to us." At that time in the Joseon dynasty Soybean curd was called "*dupo*," and there was a position called "*Pojang*," or one who made "*dubu*" in the royal palace.

King Sejong, Governing with Soy

A statement indicating the frequency of soybean usage in *Annals of the Joseon Dynasty* purports that "Sejong governed with soybeans." Soybeans are mentioned in the text in 566 cases spanning the 32 years of King Sejong's rule. There are 178 mentions of soybeans during the 41 years of King Seonjo's reign, and 73 mentions during the 52-year reign of King Yeongjo. In Sejong's time there are also many more mentions of rice than during the reign of other kings. For Sejong, 893; Seonjo, 511; Yeongjo, 591. What is more surprising is the ratio of soybean mentions to rice: for Sejong, 63%; Seonjo, 35%; and Yeongjo, 12%. Sejong recognized the value of food and is seen as a king with a particular love for soybeans. Some of the very diverse ways in which Sejong used soybeans include as a congratulatory gift for a couple who bore triplets, aid food in times of famine, a source for seeds, horse fodder, wedding or funeral gifts, commerce, stipends, taxes, elephant feed, royal gifts, food for refugees or displaced persons, loans, appeasements, and more.

King Munjong (r. 1450-1452): The Salty Water Problem

In the first year of Munjong (1451) there was a council in the Royal Court regarding the production of *dubu.* Salt is required in manufacturing *dubu,* but because the salt fields were plowed by oxen the fields were not clean. It was commonly said that "especially in sacrificial rites and royal offerings, acidic water must be used for the *dubu.*" Munjong asked, "When making the *dubu,* what kind of water are you using? One person said they dissolve salt in water, another said they use sea water, and the council did not know which method was right." When making soybean curd in the Royal Court at that time, they wondered if they should keep using the unhygienic salty water. Minerals such as calcium and magnesium cause the protein in soymilk to coagulate. Even when adding a little salt to the vinegar in the acidic water the bean curd still coagulates. There are things we can learn even from a 600-year-old debate!

Gang Hui-maeng (1424-1483): A Record of Green-kernel Black Beans and Winding Snoutbeans

Within official Joseon dynasty documents the word used for soybeans was *suk* or perhaps *du*. Among the people, however, *tae* was in common usage, as in green kernel black beans (*seori tae*) and winding snoutbeans (*seomok tae*). During the reign of King Seongjong, civil servant Gang Hui-maeng spent some of his retirement in Geumyang village (present-day Gwacheon), Gyeonggi Province, writing *Geumyang Miscellany* (1492). Gang also farmed and got along well with other farmers in the area. In his book he recorded colloquial words exactly as he heard them in life, introducing soybean name ending in *tae* (太) for the first time in print. Since then, agricultural books used both *du* and *tae* to refer to soybeans. Even today, *kong*, *du*, and *tae* are still used extensively, depending on the age of the person, the region they are from, and common usage. *Baek tae*, *heuin kong*, *norang kong*, *jang kong*, *meju kong*, *geumeun kong*, *seomok tae*, *seori tae*, and others all fall under the umbrella of large *soybean* (*daedu*).

Heo Yeop (1517-1580): Chodang Dubu

In the mid-16th century Chodang Heo Yeop was an official living in Gangneung. At that time there was a fresh spring in the front yard of the government building. Heo made soybean curd from the well water, seasoning it with sea water. Soon his bean curd gained a reputation for being quite delicious. Heo attached his penname, Chodang, to the product and is said to have earned a significant amount by selling it. Where the spring used to be is now

Chodang district in Gangneung City, and there is a stele that pays tribute to Heo Yeop. Heo Yeop is the father of Heo Gyun and Heo Nanseolheon (see below).

Tokugawa Ieyasu (1543-1616): Hatcho Miso

In Korea folks made their own soybean sauce, but in Japan soybean sauce was made on the village level, or even on the shogunate level. Typically Japanese miso is made with soybeans and half rice or barley, but there is a specialty item currently made in Nagoya known as hatcho miso, which is uniquely made purely of soybeans and aged two or more years. In Japan hatcho miso is prized for its deep flavor and distinctive aroma. The term "hatcho miso" derives from the distance away from Okazaki Castle where the sauce was manufactured: the distance measured eight cho, or hatcho, away (one cho = 108 meters). The factory in which hatcho miso is made is said to have existed since 1337. As Tokugawa Ieyasu was born in the Okazaki Castle in 1542, he would have eaten hatcho miso. In fact, it was said to be his favorite kind of miso.[1)]

Yi Sun-sin (1545-1598): *War Diary* "Octopus Soup"

In *War Diary*, written by Admiral Yi Sun-sin, there is mention of "octopus soup" (*yeonpotang*), which actually referred to silken bean curd soup. Nowadays *yeonpotang* refers to a clear-broth soup with lightly parboiled octopus, but in the Joseon dynasty it simply meant "clear broth soup." When Ming army reinforcements arrived in Joseon, soybean curd was an important army supply food. For example, members of the central army were paid with "meat + *dubu* +

1) http://yamasa.org/japan/english/destinations/aichi/hatcho_miso.html, accessed on 6/13/2016.

vegetables + salted fish + rice + alcohol." Each public official received "meat + *dubu* + vegetables + rice," and soldiers were given "*dubu* + salted fish + shrimp + rice." The one menu item common to all these provisions is soybean curd.

Heo Gyun (1569-1618): Expert Dubu Makers

In 1611 Heo Gyun wrote *Domundaejak*, Joseon's first book of food criticism. Heo Gyun penned this book while he was in exile in Hamyeol, North Jeolla Province. When he was limited to eating only the foods that passed through his place of exile, he remembered the delicious foods he used to enjoy and began writing them down one by one. Under the heading "*dubu*," he wrote, "The people living outside Jangui Gate made *dubu* so soft and silken it defies description." The "*domun*" part of his book title refers to the door of a butcher shop, and "*daejak*" means chewing loudly. Thus, the implied meaning of *Domundaejak* is "smacking one's lips at the butcher's door, thinking of the meat one cannot now eat."

Louis XIV (1638-1715): Japanese Soy Sauce

When Catherine de Medici married Henri II in the 16th century, France began to make a name for itself in cooking. When Italian cooking was grafted onto French cooking, it is said that France reached its culinary prime. When Louis XIV and Louis XV raised themselves to the level of gourmand, the cuisine was raised to a level that would satisfy them. This was also the period in which Japanese soy sauce had been introduced to the Netherlands through the East India Company. In the Netherlands, soy sauce was generally used to make salad dressing, which consisted of soy sauce, lemon juice, crushed garlic, hot pepper flakes, olive oil, sugar, and vinegar. When Louis XIV came to power, soy sauce was one of the darling spices of the nobility. Japanese soy sauce was the first soybean food product introduced to the Western world.

Yi Ik (1681-1763): Three Beans Council

In the early days of the Practical Learning philosophy group (*silhak*), Seongho Yi Ik wrote in his *Seongho Miscellany* that "Soybeans provide the most strength to people in their daily lives." At 72 years of age, Yi Ik began cultivating soybeans on his own and created the "Three Beans Council." At the Three Beans Council, members enjoyed eating three dishes made with soybeans: soybean porridge, soybean sprouts, and *doenjang* (fermented soybean paste). Relatives would all meet to eat and to take turns reading good books aloud and discussing and studying the affairs of daily living. Yi Ik took an unusual interest in the policies and crops that were used to bring relief to families whose lives were devastated during the frequent seasons of famine in Gyeonggi Province. Among famine relief crops, Yi Ik thought soybeans were an excellent alternative and he proactively lobbied for their use. He even wrote a book about the topic called *Gwakurok*, or *Record of Concern for the Underprivileged*. Yi Ik can be attributed with being the first Joseon scholar to help the kingdom recognize the full value of soybeans.

King Yeongjo (1694-1776): The King who Loved *Gochujang*

Yeongjo, the king who reigned longest in the Joseon dynasty (52 years), was also the lone fastidious eater. According to the diary of the Royal Secretariat, there is a mention of *gochujang* (red pepper paste) on July 24, 1749: "Long ago, the royal meal always had to include something salty and something fiery. Nowadays, kings prefer things like glutinous rice with hot peppers as well as *gochujang*." Yeongjo judged that the manufacture of *gochujang* in the homes of nobles was insufficient, and that places like medical centers should also produce the paste. He said that "If there are pine mushrooms, raw abalone, baby pheasant, and *gochujang*, these four, then one will always eat well and the flavors will never get old."

Benjamin Franklin (1706-1790): Bringing Tofu to America

The first person to bring soybeans to America was a sailor named Samuel Bowen who had gone to China. However, because of the vastness of the land and the lack of developed communications technology, when something happened in a given state, others often never heard about it. Such was the case with Samuel Bowen. However, when Benjamin Franklin, one of the fathers of the new nation, was ambassador to France he acquired soybeans and sent them to a botanical garden (Bartram's Garden) in Philadelphia. Franklin even enclosed a letter informing the growers what to be careful of when planting soybeans. He added, "If you grind the soybeans and solidify them using salty water, you get a white, cheese- like product called 'tofu.'" Americans consider Benjamin Franklin to be the person who brought soybeans and tofu to America.

Park Ji-won (1737-1805): Making *Gochujang*

Master Yeonam's Letter Album is a volume containing 32 letters written by Park Ji-won during the time that he held office as a government official. Most of these letters were written to his sons, especially his eldest. He became a widower early on, and as he learned how to fend for himself with housekeeping, he even began making his own *gochujang*. "In the spare moments between working for the county, when I was at leisure, I would spontaneously sit down to write or perhaps take out a book of the old masters and work on my calligraphy; suddenly the day was gone and what work had I done? Over a span of four years I read *Classifications* intently.... Put *gochujang* into a small pot and put it in the *sarangbang* (the men's study), where if you place it next to your rice every time you eat, it will be a good thing. I made some myself, but it is not fully ripe yet."

Jeong Yag-yong (1762-1836): Fried Dubu Society

Among Jeong Yag-yong's books researching the origins of words, *Aeongakbi*, volume one contains the following: "Near all royal tombs is a monastery where soybean curd is offered and which is called a ceremonial *dubu*-making temple." Buddhist temples where *dubu*-making was taught played a large role in the nutritional science behind soybean curd being part of *sosik sochan*, or "eat little food, do not eat dishes mixing meat and fish." Also, in Jeong's *Dasan Poetry Collection* he skillfully depicts everyone assembled for and enjoying the *yeonpohoi* (fried *dobu* society).

Each of five houses supply one chicken
Soybeans are ground to make *dubu*, which is placed in a basket
The *dubu* is cut into regular square dice
Cut the ritual threads just the length of your longest finger
Mix and add mulberry and pine mushrooms
Season with fragrant black pepper and manna lichen
Monks guard against killing and will not touch it, so
The youth roll up their sleeves and slice the meat themselves.

Into a cast-iron pot with no legs they place all the food and build a fire
First they boil and then simmer it, bubbles rising to the surface

Samuel Bowen (1760): Bringing Chinese Soybeans to America

In 1760 American Samuel Bowen went to China, where he had gone as a sailor with the East India Company. He visited various places throughout China, but due to political issues between England and China, he was detained for five years. During that time, Bowen seems to have paid close attention to how the Chinese eat soybeans. In 1765 he returned to his hometown of Savannah, Georgia, carrying a sack of soybeans. He planted the beans, and the next year made soy sauce, which he attempted to export to England. Although he obtained a patent in England for his method of making soy sauce, due to health reasons he was not able to maintain his business for long.

Chusa Kim Jeong-heui (1786-1856): *Dubu Yechan* (Great Boiled Soybean curd) and Aged Soy Sauce

Even Chusa, who is judged to be the most remarkable scholar of the 500 years of the Joseon dynasty, has a connection with soybeans. In his work *Dubu yechan*, he seeks to discover the simplicity of human happiness.

Good side dishes are *dobu*, cucumbers, and gingered sprouts
Grand meetings are couples with sons and daughters.

When Chusa was exiled to Jeju Island he wrote his wife a letter containing the following: "The soy sauce from Seoul is so salty it's bitter and gives me embarrassing intestinal trouble every day. Please, quickly somehow find someone to get some soy sauce in the country and send it, and I will try to endure the wait. If you need to obtain aged soy sauce in Seoul, please send me a little. It's useless to send, though, it if it's of poor quality."

Commodore Perry (1794-1858): Collecting Japanese Soybeans

In 1850 Americans rescued a Japanese crew from a shipwreck near San Francisco. As a symbol of their gratitude, the Japanese sailors presented their rescuers with a bag of soybeans. The seeds were sent to the New York State Agricultural Society, the Massachusetts Horticultural Society, and the US Patent Office. In 1852 Commodore Perry led a heavily armed fleet to Japan to demand they open their ports to trade. An agriculturalist with the fleet ended up collecting 1,500-2,000 different kinds of seeds and plants in Japan. In 1898 the USDA's Office of Foreign Seed and Plant Introduction was established, and by 1928, during a span of about 30 years, approximately 3,000 soybean samples were collected from Japan, China, Korea, India, and other locations.

Martinus Beijerinck (1851-1931): The Discovery of Root Nodule Bacteria

In East Asia, where soybeans have been planted for a very long time, it has been known that where soybeans are planted, the soil grows richer. Farmers did not know, however, what it was that improved the soil so well. In 1765 soybeans were first planted in America, in 1786 they could be found in a German botanical garden, and in 1790 they were cultivated in a Parisian botanical garden. Throughout this time experimental cultivation of soybeans took place. In 1888 a microbiologist from the Netherlands, M.W. Beijerinck, succeeded in cleanly dividing bacteria from the root nodules of soybean plants. He learned that root nodule bacteria are aerobic, returning nitrogen needed for healthy plant growth to the surface layer of the soil. He also learned that root nodule bacteria have many types, which differ according to the soybean or other plants that act as their host.

William Morse (1884-1959): Father of American Soy

In 1929, US Department of Agriculture employee William Morse wanted to discover the various genetic resources of soybeans and put together an "expeditionary party seeking Oriental plants." The soybean expedition went to Japan, Korea, and China, for a total of two years, almost two months of which was spent in Korea. Among the 4,578 specimens of soybean genes collected, 3,379 (73.8%) were from Korea. The fact that Morse could collect so many soybean seeds in such a short time in Korea indicates the ease with which he found a great variety of beans at village markets as he traveled around the country. From 1907-1949 Morse led the development of soybean crops on US farms until the crop became one of the mainstays of the country, thus leading to his appellation as the father of American soy. He also served three times as President of the American Soybean Association.

Henry Ford (1863-1947): King of Soy

One of the foremost industrialists of the 20th century, also known as "King of cars," Henry Ford had yet another title: "King of soy." In 1929, in the aftermath of World War I and Black Tuesday, America was seeking an alternative for next-generation foodstuffs and fuel. The research center at The Henry Ford in Michigan cultivated and tested many crops, and the crop deemed most valuable at that time was soybeans. Not only were soybeans high in protein and lipids, but there were also a variety of ways in which their high fiber content could be used. Since moisture content in soybeans is low, they are easy to preserve and can be replanted every year. On his 78th birthday Ford wore a suit and tie made out of soybean fibers. He learned how to make plastics from soybean fibers and even made a "soy car" out of soybean plastic. The inside of the car was made from soybean plastic, the paint was made from soybean oil, and even the fuel was made from soybean oil.

Jung Jae-won (1917-): Pioneer of the Soymilk Industry

Jung Jae-won (honorary president of Jung's Foods Co.) turned soybean milk into a worldwide food industry. From 1937, while he was working as a pediatric doctor, he noticed that many children did not digest breast milk or cow's milk well, and some even died due to related complications. In 1964 it was revealed that there is a precise principle behind this problem, namely the lactose contained in the milk is indigestible for many people, a condition known as lactose intolerance. Jung judged that soymilk could become an alternative to cow's milk. Within two years he acquired a patent and received authorization to produce his nutritional drink. In 1968 he purchased a vacant lot next to a hospital and built a manual production factory, gradually increasing output. In 1973 in Yongin City, Gyeonggi Province, he began mass production in the Singal Factory. The name of the product was a combination of "vegetable" and "milk," resulting in "vegemil."

Jeong Du-hwa (1918-2010): Revival of Traditional Soy Sauce

In 1970 Jeong Du-hwa began to restore sauce-crock platforms, which had begun to disappear, and to scale up the number of crocks to several hundred. He raised a banner for his work, "The mind of the people is for saving traditional sauces!" Jeong called himself "the big farmhand" and in Yangpyeong, Gyeonggi Province, built a farm he named Sujinwon for making traditional *doenjang* (fermented soybean paste). Although Jeong established the first shoe polish manufacturing company in Korea during the industrially weak 1950s, he soon bequeathed leadership to his son and reclaimed a barren wasteland, turning it into a farm whose soybeans became the sole source for his *doenjang*. Today there are hundreds and thousands of traditional sauces in production, all of which have followed on the success of the very first such farm, Sujinwon. Jeong established the guidance that traditional *doenjang* must age for at least two years, and traditional soy sauce must age for at least three years in order to be made correctly.

Kwon Tai-wan (1932-): Founding a Soybean Museum

In 1979 as Department Head at the Korea Institute of Science and Technology, Dr. Kwon Tai-wan received a grant for one million dollars from the German International Cooperation Program, which he used for Rice Processing Centers, formerly known as the Rice Drying Facility Project. In 1990 Dr. Kwon went to Germany to present the results of the project, and while there he had the opportunity to tour a bread museum. He discovered that not only was there a bread museum, but also at least ten other museums for single food items, including beer, potatoes, and asparagus. On the way home from Germany, Dr. Kwon pondered whether there could be a soybean museum in Korea–after all, Korea was originator of soybean culture. In 1984 he created and led the Korea Soybeans Society and in 2001 the Committee to Advance the Establishment of a Korean Soybean Museum. Finally, in 2014 he became the driving force behind the founding of the Soyworld Science Museum.

Kwon Shin-han (1931-): Theory of Soybeans' Place of Origin: Korean Peninsula

Dr. Kwon Shin-han, the first person in Korea to be considered a doctor of philosophy in soybean, constructed the idea of Korea as the place where soybeans originated. He based his theory on the huge genetic variety of soybeans that had been collected in Korea, including intermediate types, and the extensive spread of wild soybeans across the nation. In 1971 he presented his theory at the International Conference in Australia and caused a sensation. In order to prevent the extinction of native varieties of soybeans that had been handed down ancestrally in an unbroken line, he collected 3,000 native plants and established a genetic bank for their preservation. Due to Dr. Kwon's investigation, analysis, and evaluation of soybean characteristics, he won a Seoul City Cultural Award in Natural Sciences in 1984. In 1998 Dr. Kwon Shin-han joined and Dr. Tai-wan Kwon, Hong Eun-hui, Lee Cherl-ho and others in writing in Society Journal "Inquiry into the Validity of the Establishment of a Soybean Museum on an International Scale."[1] This provided the basis for the birth of the Soyworld Science Museum. Dr. Kwon Shin-han revealed that William Morse, dispatched by the USDA, collected over 3,000 specimens of native soybean plants between 1929 and 1931 on his journey through Korea, China, Manchuria,

1) Kwon Tai-wan, Kwon Shin-han, Lee Cherl-ho and Hong Eun-hui, The feasibility study on the establishment of a world soybean center, Korea Soybean Digest, 18(1):1-25(1998)

and Japan, and that between 1901 and 1976, 5,496 native specimens were collected, among which 3,200 are currently held and preserved at the University of Illinois. He initiated in Korea the first trial of breeding a soybean by using radiation energy.

Hong Eun-hui (1934-2008): Authority on Soybean Hybridizing

Dr. Hong Eun-hui is revered as the top authority in Korea in the field of pulse cultivation and breeding, including red beans and mung beans, among others. In 1969 he became the first person in Korea to foster many kinds of soybeans through hybridization, including *gwanggyo, bongui, deokyu, gangnim, Hwang Keum*, and jangyeop. Dr. Hong wrote 134 research papers, including "Characteristics and quantifiable interpretation of the development of Korea's *Manpa* soybean," as well as seven books. In 1986 he received a Green Ribbon Medal for his life's work with soybeans. In 1995 as a retirement memorial Hong published a monumental volume on the major pulse varieties in Korea, complete with photographs.

Steven Kwon (1947-): Planting Soybeans in Afghanistan

As Dr. Steven Kwon took charge of the development of nutritional foods for global food company Nestlé, he was sent to Afghanistan, where he discovered a high rate of mortality among mothers and infants. The number one cause of this problem was poor nutrition. He urgently sent soymilk for the children to drink. Within three months there was a sharp improvement in their nutritional levels, and the infant mortality rate decreased markedly. Dr. Kwon created a nonprofit called Nutrition and Education International to provide more formal and systematical support. Through this organization he invited Afghans to plant soybeans directly. It has been seven years since villages that had grown poppies now grow soybeans, and in each of Afghanistan's 34 provinces the cultivation of soybeans has proliferated, in a great win for the country. Now there is even an in-country soymilk factory under construction. Strong efforts are being exerted to solve the problem of malnutrition in Afghanistan.

Kenneth Setchell (1949-): Research on Physiological Functionality

Dr. Kenneth Setchell, a professor at the University of Cincinnati College of Medicine, studied steroid hormones and is famous for his work on cases of genetic defects in cholesterol and bile acid synthesis. In this process he became interested in the isoflavones in soybeans and focused his research on the ingestion of soybeans to reduce hormone-related disease risks. He has become the foremost expert in the field of isoflavones. His ties with Korea are deep: in November 1998 he attended the International Soybean Symposium jointly hosted at the Silla Hotel in Seoul by the Korea Soybean Society and Jung's Foods Company. He said, "The reason Asians tend to have a lower risk of disease compared to Westerners is because they intake mainly vegetable foodstuffs, which contain physiologically active matter," the most representative of these being the isoflavones in soybeans.

The Culture of Soybeans and Sauce

Chapter 02

1. The origin of soybean food products
2. Traditionally processed soybean products
3. Sauce (醬) culture in Korea

1. The Origin of Soybean Food Products

The Origin of Soybean Sauce and Paste (醬)

In ancient literary works, the first appearance of *meju* (dried, soybean fermention starter), called *si* (豉), is found in *Jijiupian*, a Chinese primer and proto-dictionary written during China's Han dynasty (206 BC-208 AD). A blurb in one of the chapters mentions that a man became wealthy by making *si*. Soybeans are said to have been brought to Qi, a state of the Zhou kingdom, from the southern Manchuria region in about 700 BC. During the Han dynasty the word *dujang* (豆醬) had entered into general usage in Chinese society.[1] In old Chinese documents the word *si* is considered to have come from a foreign tongue, and in *Xin Tangshu* (A New History of the Tang), *si* is recorded as being a specialty product of the Balhae region, a medieval kingdom encompassing northern Korea and southern Manchuria. Upon reading in *San kuo zhi*

1) Lee Cherl-ho and Kwon Tai-wan, History of the use of soybeans, *Kong* (Soybeans), Korea University Press, Seoul, 3-44(2005)

(Records of the Three Kingdoms)'s *Book of Wei* in the section containing the biography of the Eastern Yi tribe that "The Goguryeo people produce outstanding fermented sauces," it is possible that the Chinese people considered Goguryeo to be an advanced kingdom with excellent fermentation technology. At this time, the scent of *meju* was considered the scent of "Goryeo people." This indicates that the Chinese people seldom ate *meju* products, while the people of Goryeo cooked with *meju* products regularly and thus had the scent about them.

Today the word *jang* (醬, sauce) generally represents what used to be called *dujang* (豆醬, fermented soybean sauce and paste), which in ancient China for a long time meant *yukjang*, or a sauce made from meats. Birds or beasts were hunted and preserved in salt to make a *jang*. Gradually the production of *yukjang* declined and soy sauce and *doenjang* (fermented soybean paste) became more broadly used, so that *jang* came to be the term designating *dujang*, or sauce made with soybeans.[1)]

According to *Samguk sagi* (*History of the Three Kingdoms*), a record of the Korean Goguryeo, Baekjae, and Silla kingdoms, when King Sinmun (638) marries Kim Heum-un's daughter, the goods that are sent as wedding gifts include rice, alcohol, oil, honey, beef jerky, pickled seafood (*jeot*), and *si*, among other things, totaling 135 cartfuls. During the reign of Goryeo's King Sukjong (1103), Sun Mu from China's Song dynasty wrote a book of Korean vocabulary called *Jilinlei shi* (A work on the Korean language), in which

---

1) Lee Sung-woo, *Hanguk sikmunhwasa* (Cultural history of Korean cuisine), Gyomun Publishers, Seoul (1984)

it is noted that "*Jang* means *miljo*." *Miljo* is a variation of the word *meju*. Depending on the era, the word has been written variously as *myeojo*, *mejo*, and *miju*. One of Japan's Shōsōin documents (739) and the early tenth-century text *Wamyō ruijushō* (a dictionary of Chinese characters) state that *jang* from Goguryeo came to Japan and was called *miso*. In a work penned in 1717 called *Donga* (East Asia), it is written that "The Goguryeo sauce *maljang* crossed over to Japan and is called, according to the Goguyeo dialect, *miso*." The term *jang* first appears in East Asia in the 1st-century text *Zhou li* (Rites of Zhou) as meaning "*hae* (醢, or *yukjang*, fermented meat sauce) or *hye* (醯, another type of *yukjang*)." In *Shuowen Jiezi* (an early Chinese dictionary from about 100 AD) it says "*Jang* is *hae* and is a sauce made from meat and aged alcohol." After the Han dynasty, *dujang* in East Asia gradually developed and became differentiated into *si*, *maljang*, and various *jang* made from grains. A Japanese scholar posits that from ancient times the main sauce in China was *yukjang*, in Japan *eojang* (fermented fish paste), and in Korea *dujang*.

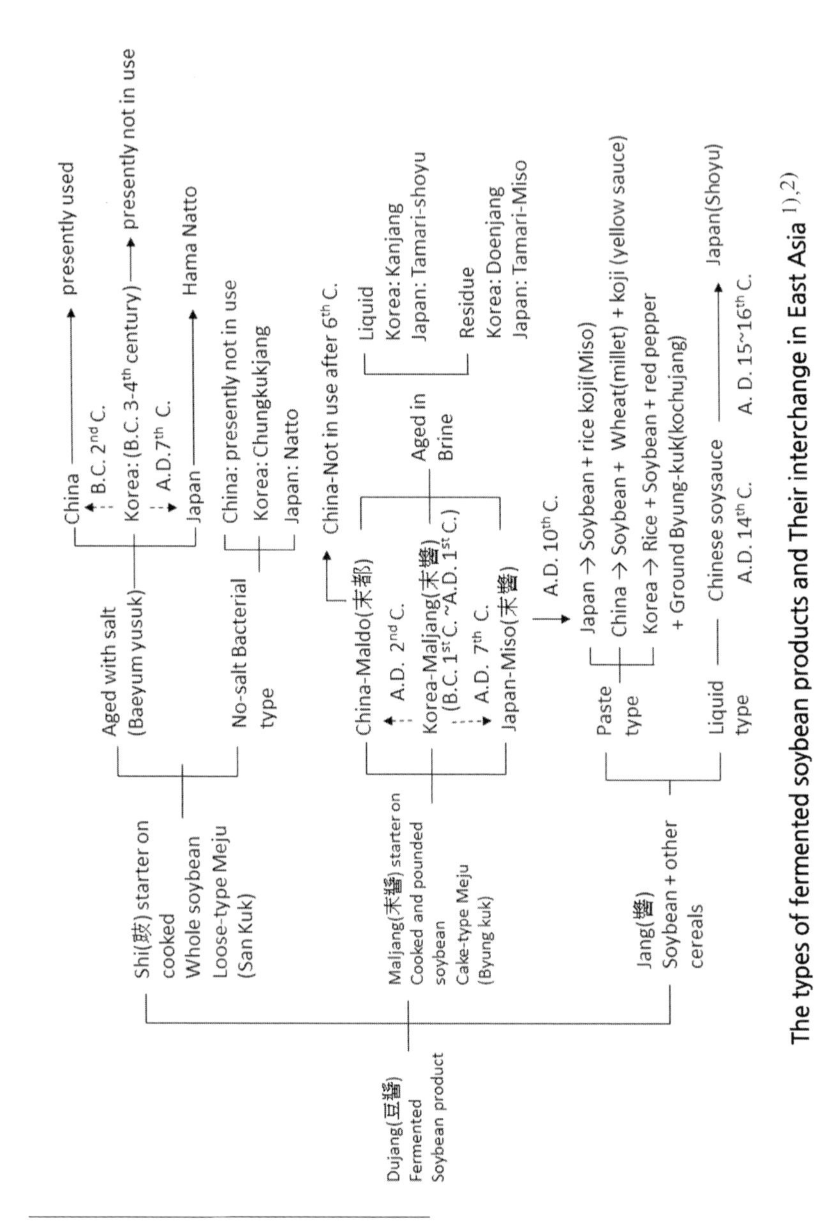

The types of fermented soybean products and Their interchange in East Asia [1),2)]

1) Lee Sung-woo, A study on the origin and interchange of Dujang (also known as soybean sauce) in ancient East Asia, J. Korean Society of Dietary Culture, 5(3):313-316(1990)

2) Lee Cherl-ho, Food Biotechnology, in Food Science and Technology, Ed. by G. Campbell-Platt, Blackwell Publishing, 85-114(2009)

It is difficult to determine when *jang* began to be made in Korea, but there is a record of *jang* in *Samguk sagi* (*History of the Three Kingdoms*, 1145), and in Japanese documents from the early 700s Goryeo *jang* is mentioned. Thus it can be confirmed in the historical record that Koreans have been making *jang* for at least 1400 years. In the 9th year of King Myeongjong (1554) in the Joseon dynasty the first technical document regarding *jang*, *Guhwang chwalyo* (Preparing for famine), was published by Yi Taek. Previously, however, King Sejong's Royal Physician Jeon Sun-ui introduced the making of *jang*, including *maljang*, *hapjang*, *ganjang* (soy sauce) and 10 more for a total of 13 *jang* recipes in his *Sanga yorok*,[1] the first cookbook in Korea.

Early Joseon Jang Varieties and Related Words

Sasichanyocho	Jang, pojang, jeupjeo
Danjongsillok	Jinjang
Hunmongjahoe	Jang, cheomjang, danjang, jangyu, si, jeonguksi, dusi
Guhwangchwalyo	Pojang, taegakjang, taeyeopjang, dosiljang, cheongjang, maljang
Miamilgicho	Ganjang, gamjang, taejang, maljang, jeupjeo
Swaemirok	Gamjang, ganjang, jeupjang, pojang, bijijang, nanjang, maljang
Donguibogam	Jang, dujang, somaekjang, yukjang, eojang, si, hwang-jeung

1) In about 1450 the Royal Physician Jeon Sun-ui wrote the oldest cookbook in Korea, containing 229 recipes.

Etymology of the word *jang*

The soy sauce and doenjang endemic to Korea are made by taking soybeans and salt as the main ingredients and boiling the beans, which is the start of meju (fermented soybean cakes). After meju is placed in salty water to ferment, soy sauce is filtered out, and the residue is consumed as doenjang (fermented soybean paste). The word for soy sauce in Korean is ganjang. The "gan" in that word means "salty," and in the word doenjang, "doen" means "hard." ("Jang," of course, means "sauce.") In the 1809 text Gyuhapchongseo (a book for the daily tasks of housewives) soy sauce is expressed as jiryeong, which in the Seoul dialect becomes jireom. The origin of this word is as yet unknown, but in 1527 a work called Hunmongjahoe, which records the meaning and sounds of 3,360 Chinese characters in the recently created native Korean alphabet, the archaic word ganjyang is used. It is thought that soy sauce became more widely used due to its high content of protein and amino acids as Buddhism, with its restriction on eating meat, was disseminated throughout the land.[1)]

The Origin of Soybean Sprouts

East Asian peoples began cultivating soybeans in about 2000 BC. It is highly probable that they knew about bean sprouts from water touching the beans causing them to sprout, but it is not clear when they began to consume bean sprouts as food. The chapter on soybeans in *Hyangyak gugeupbang*, a book on native medicines (1236) produced during the Goryeo dynasty in the time of King Kojong, introduces soybeans sprouts as medicine.[2)] This book appears before China's Yuan dynasty (1271-1368) work *Jujia biyong* (Domestic necessities), in which mung bean sprouts are mentioned and called

1) Lee Cherl-ho and Kwon Tai-wan, Introduction to Korean Food Science, Korea University Press, Seoul(2003).

2) Raw soybeans are sprouted and then dried in the sun.

duachae (in Korean pronunciation). In Korea soybean sprouts came to be called "soybean oil." This is similar to the current usage of "malt oil," which indicates the sprouting and drying of barley to make malt. The phrase "soybean oil" was used until close to the end of the Joseon dynasty, about the mid-nineteenth century. According to Chang Ji-hyun[1] it is possible that bean sprouts were cultivated before tofu was manufactured, which would date bean sprout cultivation back to the Three Kingdoms period (about 37 BC-935 AD).

The Origin of Soybean Milk

Based on remains found near millstones at middle and late Neolithic sites, it seems that soybean milk is ancient. It seems that before tofu was made people were using soymilk in their diets. The name *duyu* (soybean milk) in native Korean is *kongguk* (soybean broth), which in Chinese characters is pronounced *dujeup* (soybean juice). In Korea *dujeup* culture blossomed in tandem with *dubu* (tofu) culture, and the words *dubujeup* and *dujeup* were common throughout the Goryeo dynasty (918-1392). At the beginning of the Joseon dynasty (1392-1897) the word *tae* (太) began to be used for *daedu* (soybeans), leading to usage of *taejeup* and *taepojeup*. Towards the end of Joseon the word *duyu* came into existence and is still used today.[2]

The Origin of Soybean Curd

From long ago in places like Korea, China, and Japan soybean curd was prepared and ingested as a protein product, thus holding an

1) Chang Ji-hyun, Studies on the cooking and processing of Korean traditional soybean foods, Suhaksa(1993)

2) Son Heon-su, The history and present state of soymilk and soybean curd manufacture, *Kong* (Soybeans), Korea University Press, Seoul, 313-364(2005)

important position in the diet. In academia it is suggested that nomads like the Eastern Yi tribes had been used to milk products such as yogurt and cheese, and as they turned to a more settled lifestyle they began to make bean curd. In China, the legend of Han dynasty King of Huainan Liu An (178-122 BC) tells of him crushing up soybeans for his ailing mother in his book *Wanbishu* (Alchemical skills). By virtue of this story, Liu An is theorized to be the father of soybean curd. This inference gains credibility when juxtaposed against a painting depicting soybean curd manufacturing that was discovered in a Han dynasty tomb. However, there are those who theorize that a mural in a burial mound from 169 BC in Hunan old tamb showing a similar manufacturing scene actually depicts people making alcohol rather than bean curd.[1)] soybean curd itself is first introduced in writing in China's *Qingyilu* (An investigation of diverse curiosities) in 960: "To live a simple and frugal life, it is recommended to eat soybean curd rather than meat." Food scholars use this passage as a basis for their opinion that bean curd was made from the late Tang dynasty. It is unclear in what era the introduction of soybean curd manufacture began in Korea. In his book *Joseon sangsik* (Dialogue on common knowledge in Korea), early modern writer Yukdang Choi Nam-seon attributes the time to the Tang dynasty (618-907). The first Korean record of soybean curd appears in late Goryeo Neo-Confucian scholar Yi Saek's collected works, *Mogeunjip* (Collected Works of Mogeun) in a poem called "Daesagu *dubu* naehyang" (Bringing *dubu* to an old man): "Soupy greens get old / But dubu brings new flavor / Great food when you have no teeth / Just right for aging bodies."

1) Choi Duk-kyung, Reexamination of the origin of soybean and the distribution of soybean sauce, soybean malt and Sundubu : from the documentary and excavated data from the Ancient China, Yeoksaminsokhak No. 30, Hankook Yeoksaminsokhakhoi, 363-427(2009)

2. Traditionally Processed Soybean Products

In *Seongho saseol* (*Seongho Miscellany*), Neo-Confucian scholar Seongho Yi Ik writes, "Soybeans are one of the Five Sacred Grains, but people do not appreciate their value. In my opinion, soybeans have the most power of any grain to save people's lives. If the poor can obtain soybeans and eat them, it is the only thing that will keep their lives intact." He further suggested that planting soybeans in areas around homes such as on the ridges between rice paddies and fields would be the easiest way to bring to the people healthy food products like *kongbap* (cooked rice with soybeans), *dubu* (soybean curd), *kongnamul* (seasoned soybean sprouts), and *jang* (fermented sauces).

Dubu

The fact that *dubu* has a long history and tradition, having been passed down for generations in Korea, is probably due to its uniquely light and clean taste and texture. In order to make *dubu*, first high-quality domestic soybeans must be selected and then soaked. In the summer they should soak 7-8 hours, and in the winter about 24 hours. Well-soaked beans then go to the millstone. Gradually the soaking water is poured out and replaced. The ratio of beans to water should be about two parts beans to three parts water. Ground beans are stirred continually so they do not burn as they simmer. Boiled soymilk is then squeezed through a cheesecloth or muslin. This filters the dregs from the soybean water (soymilk), the latter of which is boiled again. *Gansu* is stirred in so the curd will slowly coagulate; at this stage the very soft *dubu* is known as silken *sundubu*. *Gansu* is salty water: a straw bag of salt is propped up and water is dripped through it; the water that drips out the bottom is caught and used for the bean curd. Cotton cloth is spread in a square mold that has small holes poked into it, and the fresh, soft curd is placed inside. The cloth is wrapped around the curd, the box closed with a lid, and a heavy stone placed on top. In about 15-20 minutes enough water will be removed for a lovely, complete *dubu*. Towards the end of this time, hands are used to gently press down in order to adjust the time and make an even smoother, more delicious bean curd.

Soybean Sprouts

Raw soybeans do not contain vitamin C, but when sprouted, a great quantity of vitamin C is formed. Seongho Yi Ik has said, "If the poor would grind up soybeans and chop the sprouts they could make a good porridge that would amply fill their bellies." Soybean sprouts have been an important food and one of long tradition in Korea, with dishes like seasoned sprouts and bean sprout soup. In order to grow soybean sprouts at home, first prepare a container full of water. Submerge the soybeans completely in the water and leave them to soak 3-4 hours. If they are left for too long they will lack the energy to support sprouts, so do not soak them too long. Next, in the course of one day divide them 5-6 times, giving water each time. Give enough water to wash away any organic matter left on the sprouts. If the temperature of the water is high, the sprouts will grow faster, but they will easily rot; if the water temperature is too low they will grow too slowly. It is best to give them lukewarm water. Soybeans must be raised with no light, but rather than seal them into a container, cover them with a dark cloth. Once the soybeans begin to sprout, they will be ready to eat in 5-6 days.

Doenjang (fermented soybean paste)

Since *doenjang* and soy sauce are used as basic ingredients that bring out the desired salty seasoning in food, making them has been considered as important an annual event as making *kimchi*. The *meju* (cakes of dried soybean fermentation starter bean paste) used to make sauces is begun by boiling and pounding soybeans in about the 10th to 12th lunar month, then fermenting. Sauces are made the next year at the onset of spring, just as the cold begins to lift. Neatly trimmed *meju* are placed in salted water to soak. They sit in a ceramic crock with the lid on for about four days. On a sunny day the lid is removed in the morning to allow the sun to shine down on the sauce all day, after which the lid is replaced. Covering the mouth of the crock with cotton or mesh cloth prevents foreign materials from falling into it. In about 40-60 days the sauce will ripen enough to achieve separation of the juice from the *meju*. The juice becomes soy sauce, and the *meju* becomes *doenjang*. Sauce separation depends upon storage time and differs somewhat from region to region. The sauce must be removed from the crock carefully in order to keep the *meju* as intact as possible. The fragments of *meju* left on the bottom of the crock are propped up in a sieve and dried. The dried *meju* is again mixed with salt and then packed tightly into another crock. The flavor of the sauce depends on the *meju* and salinity, exposure to the sun, and the amount of time aged. If the concentration of salt in the water is too low, the contents might spoil during the ripening or storage stages. If it is too salty, the microorganisms present in fermentation will be suppressed and the sauce will lose its flavor. If there is much more salt water than *meju*, there will be too much soy sauce and it will have a weak flavor; if there is less water, there will be little soy sauce with a strong flavor. In order to make good soy sauce, then, there should be little water and plenty of *meju*. The typical ratio of *meju* to salt to water is 1:1:3-4.[1),2)]

Types of *Doenjang*

- *Makdoenjang*: The by-product of removing the soy sauce. Typically referred to as *doenjang*.
- *Tojang*: *Doenjang* made with *meju* only-the soy sauce is not removed but is left to age at room temperature for a long time.
- *Makjang*: Raw *meju* is ground into a powder and made soft and wet by mixing it with salt water and then left to ripen. North of the central regions of South Korea this sauce is not steeped, but in Gangwon and Gyeongsang Provinces it is steeped well. In Chungcheong Province *meju* powder and *gochu* (red pepper) powder are mixed and seasoned with salt, then placed on barley rice and steeped, while in Gyeongsang Province soybeans and non-glutinous rice are mixed in the making of *meju*.
- *Dambbukjang*: *Meju* is finely ground and mixed with *gochu* powder, then covered with water and left to stand overnight, at which point it is seasoned with soy sauce and salt. In Chungcheong Province *meju* powder and *gochu* powder are mixed into the water strained from bean curd, and in Hwanghae Province barley is made into a porridge called *juk* and then mixed with *meju* and *gochu* powders and salt.
- *Bbamjang*: For the purpose of making *doenjang*, *meju* is coarsely ground, boiled in salty water, then stored in cooled water. This is typically made in Gyeongsang Province.
- *Bbagaejang*: A sauce made in Chungcheong Province in which *meju* powder is mixed with the water soybeans were boiled in, then *gochu* powder and salt are mixed in.
- *Garujang*: In Gangwon Province raw barley is ground up, steamed, and mixed

1) Han Bok-nyo and Han Bok-jin, Sauces of the Korean people, Kyemunsa, Seoul, (2013)
2) Sin Dong-hwa and Lee Hyo-ji, Fermented soybean foods, *Kong* (Soybeans), Korea University Press, Seoul, 365-406(2005)

with *meju* powder, then boiled. Cool, salted water is poured over it, and it is seasoned and steeped.

- *Borijang*: Barley is boiled and brewed and then ground to a powder. It is mixed at a one to one ratio with *meju* powder and steeped. This sauce is made on Jeju Island.
- *Cheongtaejang*: Soybeans that have not been dried are steamed in an earthenware steamer and formed into a rice cake, then covered with soybean leaves and brewed. The *cheongtae meju* is brewed in a hot place and seasoned with sun-dried *gochu* (Korean red peppers). The soybean leaves are used to more readily break down the bacteria.
- *Patjang*: Red beans are boiled, balled up to brew, and then mixed with soybeans to steep.

The reason boiling doenjang for a long time makes it tasty

Even when *meju* is turned into a well-brewed sauce, the proteins do not break down completely, and many peptides remain. The longer *doenjang* is boiled in an earthenware pot, the more the peptides will break down into amino acids, which provide a better flavor. However, if boiling *doenjang* for a long time using an improved *meju* for *doenjang* stew, the amino acids will permeate the vegetables, meat, and tofu in the stew and the broth will lighten. Thus commercial *doenjang* made with improved *meju* can be boiled for a short time and still have good flavor, like that of *doenjang* boiled for a long time in the traditional way.

Soy Sauce

A book written in 1766, *Jeungbo sallim gyeongje* (Revised farm management) states "*Jang* (sauce) contains the same Chinese character as the *jang* (奬) that means jangsu(奬水) and thus is considered the basis of all flavors." *Jang* is considered the first of all foods because it seasons so many of them. It was always imperative that households make *jang*. If the flavor of one's *jang* was not good, it would not matter how good the vegetables and meat were; no good food could be made in that house. On the other hand, if people living in the countryside could not obtain meat easily, good *jang* would mean never worrying about the taste of one's side dishes (*banchan*). As a foundational flavoring, soy sauce goes well with salty, sweet, savory, and other flavors and has a unique taste and smell. Soy sauce is made by boiling down the juice from *meju* steeping in salty water. Soy sauce can be boiled down and aged as is. A darker flavor is achieved when *doenjang* and soy sauce are divided and soy sauce is boiled down separately, thus preventing the soy sauce from decomposing and concentrating its flavor. Soy sauce is boiled down continually at a temperature of 80℃ for about 10-20 minutes, and as it boils the froth formed on top is skimmed away. The boiled sauce is cooled and poured into a crock, then covered with a lid. Aged soy sauce sometimes has new soy sauce mixed in to make a combination sauce.

Types of Soy Sauce

There are soy sauces of differing concentration: dark soy sauce, medium soy sauce, and light soy sauce. Each is used with different foods, depending on the degree of desired saltiness or sweetness; the color also differs. Young soy sauce, which has been aged only one or two years, is used for things like soups; medium soy sauce has been aged about three or four years and is used in stews and seasoned greens; dark soy sauce has been aged for at least five years and is used in *yaksik*, a sweetened, glutinous rice dish, and seasoned abalone, among other dishes.

"Soy Sauce Ambassador" in the time of King Seonjo (r. 1567-1608)

Long ago during national crises when the king had to be evacuated, a public official called the Soy Sauce Ambassador would travel ahead of him to the king's destination to acquire the sauces the king would eat. During the 2nd Japanese invasion King Seonjo appointed a government official from Hamgyeong Province by the last name of Sin to be the Soy Sauce Ambassador, but the officials in the Royal Court all stood up in opposition to this man. The reason for their opposition was simply that when sauces are made, they are never made on a day called *sin*, which is a homonym of that government official's last name, and due to that coincidence they feared the soy sauce would lose its flavor. There were even families bearing the surname Sin that would only make soy sauce at their in-laws' or their married daughters' homes. These anecdotes show just how highly people esteemed good soy sauce.

Gochujang

Gochujang (red pepper paste) is a sauce that harmonizes different flavors, like the sweetness that comes from sticky rice or malt, saltiness, the savory or meaty flavor of *meju* and pungent taste of red pepper. This sauce cannot be found in China or Japan, but is unique to Korea. Although *doenjang* and soy sauce date from before the Three Kingdoms period (57 BC-668 AD), *gochujang* was newly incorporated into Korean cuisine during the 16th-century Japanese invasions, when *gochu* (hot peppers) were introduced. It is thought that at first *gochu* powder was added to *doenjang* a little at a time until gradually it became the *gochujang* it is today. The meeting of the spicy peppers with traditional sauce likely became the first "fusion food" of the early Joseon dynasty. There is a clear distinction between the Korean dining table before the use of *gochu* and after the introduction of *gochu*. Today *gochu* is used in *kimchi*, *gochujang*, stews, and other side dishes and provides a much greater variety of food. *Gochujang* can be made with different starches, including glutinous rice, non-glutinous rice, or barley. *Meju* made with *gochujang* contains not only soybeans as in that which is made into soy sauce and *meju*, but also about 20% glutinous rice powder. *Gochujang* is a fermented, seasoned sauce made by mixing *meju* powder with very sticky rice, rice cake powder, or thick porridge and *gochu* powder and salt to taste. It is good with *bibimbap* (mixed rice and vegetables with or without meat), stews, spicy fish soup, raw vegetables, meat boiled in soy sauce, raw fish, and finger foods tied together with boiled scallions. It is seen as a necessity for dishes like fish boiled in soy sauce and fish stews because it removes the fishy smell. As in *yak gochujang* (see below), stir-frying meat in the sauce has become a favorite side dish.

Types of *Gochujang*

- *Borigochujang*: Often made in Chungcheong Province, raw barley is rinsed and ground into a powder, then steamed in an earthenware steamer. It is then mixed with cool water and placed in the steamer again, then set in a hot room to brew. Once white mold has formed, *gochu* powder and *meju* powder are mixed in, and it is salted to taste and placed in a crock to steep. The amount of barley is about two *mal* (or about 36 liters) to 10 *geun* (about 6000 grams) of *gochu*. One characteristic of *borigochujang* is that it does not use malt.
- *Susu gochujang*: Porridge is made from salted water and sorghum powder, into which *meju* powder, malt powder, and *gochu* powder are mixed. It is salted to taste and steeped.
- *Mugeori gochujang*: The remnants after making *meju* powder are mixed with barley powder, malt powder, and *gochu* powder. This *gochujang* is mainly used in stews and has a sweet-and-sour flavor.

- *Yak gochujang*: Meat is finely minced, seasoned, and fried in a pan with oil, later adding *gochujang*, green onions, ginger, and sugar. It is even more delicious when pine nuts are added to the cooled mixture.
- *Pat gochujang*: Non-glutinous, unshaped rice cakes are made, then soybeans and red beans are well-boiled and pounded together with the rice using a mortar and pestle until flattened, with all lumps removed.
- *Goguma gochujang*: Malt is placed on boiled sweet potatoes and brewed, then put into a hemp bag and squeezed. The resulting water is boiled down like taffy, after which *gochu* powder, *meju* powder, and salt are added.

Yi Seong-gye and *gochujang*

At the end of the Goryeo dynasty, Yi Seong-gye stopped to find the Buddhist monk Master Muhak, who was staying at Manil temple in Gurim township in the region of Sunchang, North Jeolla Province. While there he visited a farm and enjoyed a lunch that included *gochujang* so much that he later was unable to forget that delicious flavor. There is a story that after becoming king, Yi ate a dish called *chosi*, which included a sauce of hot peppers from a Chinese pepper tree mixed with *meju*. The inclusion in *Somun saseol* (a cookbook from the late Joseon dynasty) of a Sunchang *gochujang* recipe, and the mention of Sunchang *gochujang* being a regional specialty in *Gyuhapchongseo* (an encyclopedia of daily life for the homemaker, 1815), show that *gochujang* from the Sunchang region was renowned even long ago. In 1997 Sunchang received the designation “Korean Chili Paste Village,” after which a sauce research center was opened, along with the Sunchang Jang Museum and the Sunchang Jang Experience Center. Gradually this area has become known as “Jang Valley.”

Cheonggukjang

One theory about the origin of *cheonggukjang* is that it derived from the first form of soybean sauce, *si*; another theory suggests that the provisions for the Qing army during the Manchu war of 1636 included something called *sokseong* (quick) sauce, which from that time on became known as *cheonggukjang*, or *jeongukjang*–the "*jeonguk*" meaning "war nation," or "nation at war." During war or another national emergency, people cannot wait a long time for sauces to age, but must eat them as soon as possible once they have been prepared, which is why *sokseong* (quick) sauce was necessary. Often people say *cheonggukjang* comes from the time the Qing soldiers in the Manchu war loaded up cooked soybeans onto their horses to bring with them, but Dr. Lee Sung-woo stresses that "For thousands of years the Joseon people maintained a culture of *jang* (sauces), and to say that they learned about *cheonggukjang* from the nomadic hunting people known as the Jurchen [those who established the Qing dynasty] who overthrew the Ming dynasty is illogical." *Doenjang* and soy sauce are made from *meju*, and *cheonggukjang* is a high-protein food made from these original soybean materials and fermentation. Currently, commercial versions can add cultured bacteria and be manufactured in one day. Naturally-fermented *cheonggukjang* is brewed by steeping *meju* for 10-20 hours in hot water, then pouring water over it and boiling vigorously, after which it is kept warm. In the country, the cooked beans are placed in a basket and cooled to 60°C, then covered with a cotton cloth, placed in a warm room and covered with a blanket. If it stays warm for 2-3 days, the bacillus bacteria will grow and it will become fermented. Bacillus bacteria grow well at a temperature of 40-45 °C. Their job is to decrease carcinogens and absorb toxic substances to be excreted by the body. There are many bacillus bacteria in the air, but plenty more are found in rice straw, so placing rice straw between the soybeans when brewing *cheonggukjang* provides a fine fermentation.

Types of *Cheonggukjang*

- *Saeng cheonggukjang*: *Cheonggukjang* is fermented and eaten young ("*saeng*," fermented only about a day). It is often eaten rolled in salted, dried laver or eaten with *kimchi*, and can also be a main ingredient in *bibimbap* (mixed rice).
- *Cheonggukjanghwan*: *Saeng cheonggukjang* is dried, ground to a powder, and reformed into edible pellets. There are pellets made simply of *cheonggukjang*, while others are mixed with glutinous rice powder, pine needle powder, or kelp powder.
- *Cheonggukjang garu*: *Saeng cheonggukjang* is dried and ground into a powder. It is eaten mixed with yogurt, milk, or raw foods.
- *Mallin cheonggukjang*: *Saeng cheonggukjang* is dried as is. It is eaten crunchy, like peanuts.

3. Korea's Culture of Sauces

Auspicious Dates and Management

Since sauces are the foundation of all other foods, much effort has always been needed to achieve good results. People were careful to maximize their good luck on the day of making *jang* (Korean fermented sauce), so they began by choosing an auspicious day, like the 7th day of the 10th lunar month (also known as "day of the horse"). They refrained from going out starting three days before sauce-making day, and on the day of the making they even covered their mouths with Korean paper, hanji, to diffuse any potential negative energy. Strict attention was also paid to the crocks on the platform that stood outside the house. Crocks and other vessels were dusted until they shone. On sunny days the lids of sauce crocks were removed, and if it looked like rain, the lids were quickly replaced so that not even a single drop of rain would mix in with the sauce. If the flavor of a sauce should be sullied it was not only a big mistake on the part of the homemaker, but was seen as an omen that something terrible was going to happen to the family. When all was well, guests would be offered the following invitation to visit a home: "Please come and taste the wonderful flavor of our family's *jang*."

Why *Meju* is Tied with Straw

The practice of tying finished *meju* with straw lies in the fact that the bacteria *Bacillus subtilis* inhabit the straw, thus inoculating the *meju*. *Bacillus subtilis* are anaerobic bacteria that like high temperatures, the optimum being about 40℃. When *meju* comes into contact with the air, mold grows on its outer walls, and bacteria that include *Bacillus subtilis* grow on the inside. The fermentation of *meju* should be in the region of 26-30℃. If the temperature is too high, if the ultraviolet rays are too strong, or if there is insufficient moisture, the *meju* will not turn out well.

Why Charcoal and *Gochu* (red peppers) are Added

There is a traditional, magical belief as to why charcoal is added to sauce crocks when making sauce: it is thought that the holes in charcoal will trap evil spirits who come to try to ruin the flavor of the sauce. In reality, microorganisms that aid the fermentation process live in the numerous tiny holes. *Gochu*(red pepper) was also thought to have the power to drive away evil spirits due to its red color and spicy taste, but really it is the capsaicin found in hot peppers that prevents spoilage due to its sterilizing and preserving effects. People also put jujubes into sauces because their red color was supposed to repel evil spirits, but they merely sweeten the sauce.

Upside-down Socks

People used to place the paper pattern for making socks upside-down on sauce crocks, or hang them from a string, with the idea that evil spirits would fly into the socks and not be able to get out. Again, this was magical thinking behind keeping the flavor of *jang* good. There is a scientific basis for this practice, however: since millipedes hate the light reflected by the white paper, putting it on the mouth of a sauce crock would ensure that these bugs would not come around. In the region of Nonsan in South Chungcheong Province, the homemaker would place the upside-down sock pattern on the crock while shouting, "*Gguldok io*!" which means "Make my sauce sweet as honey!"

Pottery that Breathes

The pottery used for storing *jang* is called *jangdok*, or crocks selected specially for that purpose. These crocks must allow air to pass through them and so are known as pottery that breathes. Pottery that is made in midsummer is not good for *jang*, for although it is fired, the moisture will not be completely removed. It is best to purchase crocks in the spring that were made the preceding winter. Good crocks are those that emit a metallic ringing sound when tapped. It has long been known in Korea that *jang* crocks must breathe. The many grains of sand in the clay used in making the crocks create tiny holes in the walls that allow air to pass through. The air allows the sauce on the inside of the crock to age well and be well-preserved. These crocks are the best containers for the fermentation process.

Regional Variations in Pottery

Each region in Korea has its own unique crock shapes. Crocks are arrayed in rows on the crock platform in order from largest (in the back) to smallest (in the front). Generally speaking, in the warm, southern regions the crock is wide all around in order to reduce the ratio of sunlight penetration to sauce (since the sun is hotter there), while in the northern regions, crock girths are smaller in order to allow more sunlight per amount of sauce. In Gyeonggi Province (near Seoul) the opening in a crock is similar in width to its bottom, resulting in an overall slender silhouette. In Gangwon Province (in the northeast), many crocks have a shape similar to those found in the adjacent province, Gyeonggi, though some have wide mouths, and some are made smaller for easier transport in mountainous areas. In Chungcheong Province (the middle-west region) the neck of crocks is high and the bowl is a bit fuller. Jeolla Province (southwest) crocks are wider at the opening than at the bottom and have a full upper belly, rounding down from the shoulders for a plump look. In Gyeongsang Province (southeast) the torsos of crocks protrude, while the area from the shoulders to the neck constricts sharply, ending in a very small mouth. On Jeju Island, the mouth and the bottom are small but the belly protrudes, while the clay used for pottery comes from vol-

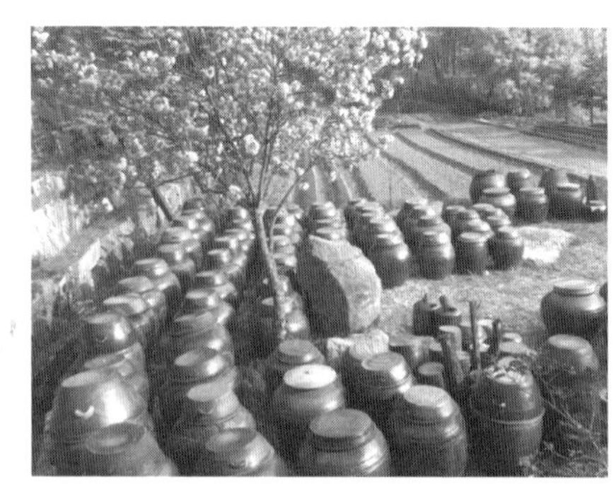

canic soil and thus is tinged with red. The distinctions found in regional crock varieties have dimmed over time due to the lively exchange of goods across regions.

Soybeans and *Jang* in Idioms

Over 80 Korean idioms have been found that contain references to soybeans or sauce. These include the following: "Your mind is in a soybean field" (your mind is far away); "It's like a bean sprout steamer" (packed like sardines); "It's like soybeans in a drought" (once in a blue moon/few and far between); "If you fall in love, soybean skins will cover your eyes" (love is blind); "Know your soybeans and then attack" (go into it with eyes wide open); "If you plant soybeans, soybeans will grow; if you plant red beans, red beans will grow" (you reap what you sow); "Like a pheasant thinking of a soybean field" (not thinking about your work, only about your personal gain); "Like bones in *dubu*" (bad luck follows you); "Eating *dubu* makes your teeth fall out" (an impossible or nonsensical situation). Following are some idioms with reference to *jang*, or sauce. "You can know a town's politics by the flavor of its drinks, and you can know the personal business of a family by the flavor of its sauces;" "If the daughter-in-law skillfully enters the family, the flavor of the house sauces improves" (the flavor of the sauces is a metaphor for the family's prosperity); "A prosperous family has sweet-flavored sauces" (in this case the sauce is not actually sweet, but rather tastes delicious). The importance placed on the flavor of a family's sauce in these idioms indicates that the women of the household had their work cut out for them over the winters. There are also many interesting idioms having to do with sauce at whose meaning one can only guess: "A household of many words has rotten sauce" (people who present a kind façade but are not actually kind); "After talking you ask for sauce" (ingratiate yourself with someone in order to ask for what you want); "The flavor of the *doenjang* gives you a glimpse of the daughter-in-law in between the sheets" (just as good *doenjang* stew must be simmered gently and in secret, so too with sexual relations); "There is more *gochujang* than rice" (there is more of the side attraction than the main event); "Fetid sauce disappears slowly" (bad things never go away quickly); "Like a dog keeping an eye on a chunk of *doenjang*" (you will fail in your work if you assign it to someone untrustworthy); "Dried out from soy sauce soup" (he's so strung out he loses his cool); "He's been marinated in soy sauce, a little vinegar won't kill him" (someone who has been through so much can handle a difficult situation).

4. Varieties of Soybean Foods

Our ancestors cultivated soybeans and made food from them for daily living, but they also used them as ceremonial foods. Professor Lee Hyo-ji has classified foods using soybeans as follows:[1)]

Staples

In rice dishes there are soybeans and soybean sprouts; in porridge, soybeans, soybean sprouts, soybean curd, *gochujang* (hot pepper paste), and *doenjang* (fermented soybean paste); in noodle dishes, soybeans, soybean powder, soybean broth, and soybean

---

1) Lee Hyo-jik, Soybean foods of Korea, *Kong* (Soybeans), Korea University Press, Seoul, 455-528(2005)

curd in dumplings, *dubu* in the dumpling dough and *dubu* inside the dumplings; in pudding, soybeans; in rice cake soup, *dubu*, and in certain circumstances steamed *dubu* is eaten straight.

Side Dishes

In soups there are *dubu*, soybean powder, soybean sprouts, *doenjang*, and *dubu* dregs; in stews, *dubu* dregs, *dubu*, soft *dubu*, *doenjang*, *cheonggukjang* (fast-fermented bean paste), *gochujang*, and soybean sprouts; in vegetables (soybean sprouts, fried soybean sprouts, soybean sprout soup, seasoned soybean sprouts), soybean sprouts; in soybeans cooked in soy sauce, soybeans; in *jorim* (fried bean curd, fresh *dubu*, *dubu* paste, and marinated *dubu*), soybeans; in meat roasted with seasonings, *dubu*, in raw vegetable dishes, *dubu*, vegetables pickled in soy sauce, soybean leaves (autumn leaves); in raw fish dishes, *dubu*, in *ssam* (dish with lettuce or other leaves used to wrap meats), soybean leaves; and in *kimchi*, soybean sprouts and soybean leaves.

Ceremonial Foods

Soybeans used in steamed rice cakes are mixed with rice flour, then the rice cakes are steamed, and the soybean powder either covers the exterior of the rice cakes or becomes part of the filling in dumpling rice cakes. In pounded rice cakes, soybean powder is used on the outside. *Doenjang* is used in pan-fried rice cakes. Soybean powder is used as a covering for boiled rice cake.

Sweets

Molded tea cakes and pan-fried biscuits use soybean flour, and puffed rice squares, pan-fried sticky rice cakes, grains mixed with taffy, and taffy all use soybeans.

Beverages

Some traditional beverages use pan-fried soybeans that are put in a strainer over which boiling water is poured, and some use pan-fried soybeans that are made into a powder and mixed with cold water and honey or sugar.

1. Rice with soybeans
2. Rice with soybean sprouts
3. Soybean sprout porridge
4. Soybean porridge
5. Soybean noodle soup
6. Dubu soup
7. Dubu dregs stew
8. Soft tofu stew
9. Cheonggukjang stew
10. Doenjang stew
11. Blended tofu squares with toppings
12. Layered rice cake
13. Rice cake encased in soybeans
14. Soybeans mixed with rice cake
15. Rice balls or discs coated in bean flour and filled with sweet bean paste
16. Plain rice cake squares coated in bean flour
17. Rice cakes
18. Doenjang rice cakes
19. Rice cake balls with fillings
20. Molded tea cakes
21. Soybeans shaped with taffy
22. 5-spice dubu
23. Dubu meat dish
24. Aromatic dubu
25. Yeoui dubu
26. Tofu in soymilk
27. Soybeans with rice and soy sauce
28. Steamed soybeans
29. Bitter gourd and tofu stir-fry
30. Dubu pizza

Various Soybean Foods

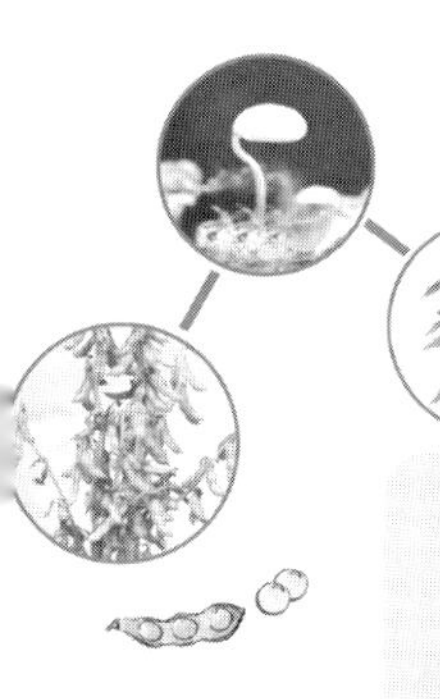

Chapter 03

The Growth and Ecology of Soybeans

1. Korean soybean varieties and varietal explanations
2. The life cycle of soybeans
3. Varietal classification by use
4. Soybean farming and cultivation methods

1. Korean Soybean Varieties and Varietal Explanations

From ancient times Koreans have considered soybeans to be one of the staple food crops along with rice. During Korea's Three Kingdoms period (57 BC-668 AD) soybeans were already being made into *meju*, or blocks of dried, fermented bean, for use in the production of sauces. In Unified Silla (668-935) and early Goryeo (918-1392) the usage of soybeans expanded to soybean curd and bean sprouts. Exactly how long ago is unknown, but soybeans penetrated deeply into the food landscape in Korea, with such varied uses as cooking them with rice, using them as rice-cake filling, and eating them in salted, dried form. The differentiation of soybean varieties improved and developed as specific varieties were matched to their usage.[1] Looking into agricultural books that describe soybean varieties, there is an early record from the Joseon dynasty, *Sasichanyocho* (Important farm facts), author unknown, that sorts soybean vari-

1) Kim Seok-dong and Lee Yeong-ho, Breeding and cultivation of soybeans, *Kong* (Soybeans), Korea University Press, Seoul, 137-216(2005)

eties broadly into the categories *cheongtae*, *heuktae*, and *hwangtae*, according to their colors: green, black, and yellow, respectively. During King Seongjong's reign (1469-1494), Gang Hui-maeng wrote *Geumyang jabnok* (a farming miscellany), which records eight varieties of soybeans with an explanation of each.

Towards the end of the Joseon dynasty a place set up for encouraging industrial models in 1906 in Suwon, Gyeonggi Province achieved a select collection of 28 recommended soybean varieties, including *Chungbukbaek*, *Buseok*, and *Jangdanbaegmok*, which have become the genetic basis of the soybeans grown in the US, the top soybean producer in the world. In addition, up until the first improved variety was introduced through cross breeding in 1969, each region of the country made use of them to produce their own choice quality soybeans.

From that time until 2014 the number of soybean varieties developed at the Rural Development Administration, in each province's Agricultural Research and Extension Services, and at universities reached about 200 varieties. Among these the favorites that were widely used for making soybean curd and sauces in the 1980s and early 90s were *Jangyeobkong* and *Hwangkeumkong*; from the mid-90s until now, *Taekwangkong* and *Daewonkong*, among others, have been predominant. As for soybean sprout cultivars, following *Junjeori*, *Seomoktae*, and *Baekjomkong* and the introduction of *Hill*, *Danyeobkong*, and others, in the late 1980s and early 1990s *Eunhakong* was widely propagated, and today the newly developed and disseminated *Pungsansamulkong* is loved for its high productivity and taste. After 1993, starting with *Geomjeongkong 1* and followed by *Seonheukkong* and *Cheongjakong*, *Cheongjakong 2* and *Cheongjakong 3* were used with salted, dried soybean, and colored soybeans were developed and propagated, responding to the demand for soybeans such as green cotyledoned black soybeans to be boiled with rice. After 1991 the demand for early and *vegetable* soybeans was fulfilled by the development and dissemination of *Keunolkong*, *Hwaseongputkong*, *Seonnok*, and *Mirang*.

From the late 1980s soybean varieties were developed to satisfy the drive for market development and the demand for high-quality soybeans adapted to the World Trade Organization system. The beany flavor unique to soybeans occurred from the lipid oxidizing enzyme lipoxygenase, so this was eliminated, making them suitable for the production of soymilk and soy ice cream. Soybean varieties with a protein content of 45% or more, 15% higher than standard soybeans such as *Danbaekkong*, and *Saedanbaek*, are currently being produced. The quality of the recently developed *Daepung* variety is somewhat lower, and cultivation stability and yield potential not as good, as the excellent *Paldalkong* and its series, *Sinpaldalkong* 2, which were developed in the late 80s–varieties that directly and indirectly received genes for high yield and particular plant type. These varieties continue to be propagated on farms.

Domestication of Wild Soybeans and Breeding of Cultivated Soybeans

Wild soybeans are either too small, or by maturity time their pods have already opened and the beans have popped out. In prehistoric times soybeans were domestigated by the ancestors of the Korean people, and now these soybeans have become today's cultivated soybeans. In order to improve cultivated soybeans, processes such as pure-line selection, the introduction of international natural resources, and cross-breeding were applied to bring forth newly developed soybeans known as "new varieties." The planting of new varieties was encouraged on a national level, and thus they became widely propagated. The business of soybean breeding in Korea began via collection of land races and pure-line selection from 1906. The first recommended variety was called *Jangdanbaekmok* was released in 1913. As the first cross bred variety, known as *Kwangkyo*, was released in 1969, development of varieties kicked into high gear. In 1980 a variety called "*Hwangkeumkong*" was developed to have large seeds with good color and gloss; this soybean was a leader of superior soybeans used for Korean *dubu* and soy sauces. In the 1990s there were *Taekwangkong* and *Daewonkong* soybeans, and in the 2000s *Daepung*, *Uram*, and other key varieties cultivated in Korea were disseminated. The key to increasing the quality of soybeans lay in developing the characteristics important to the use of each soybean plant, capitalizing on specific innate qualities of each variety.

2. The Life Cycle of Soybeans[1)]

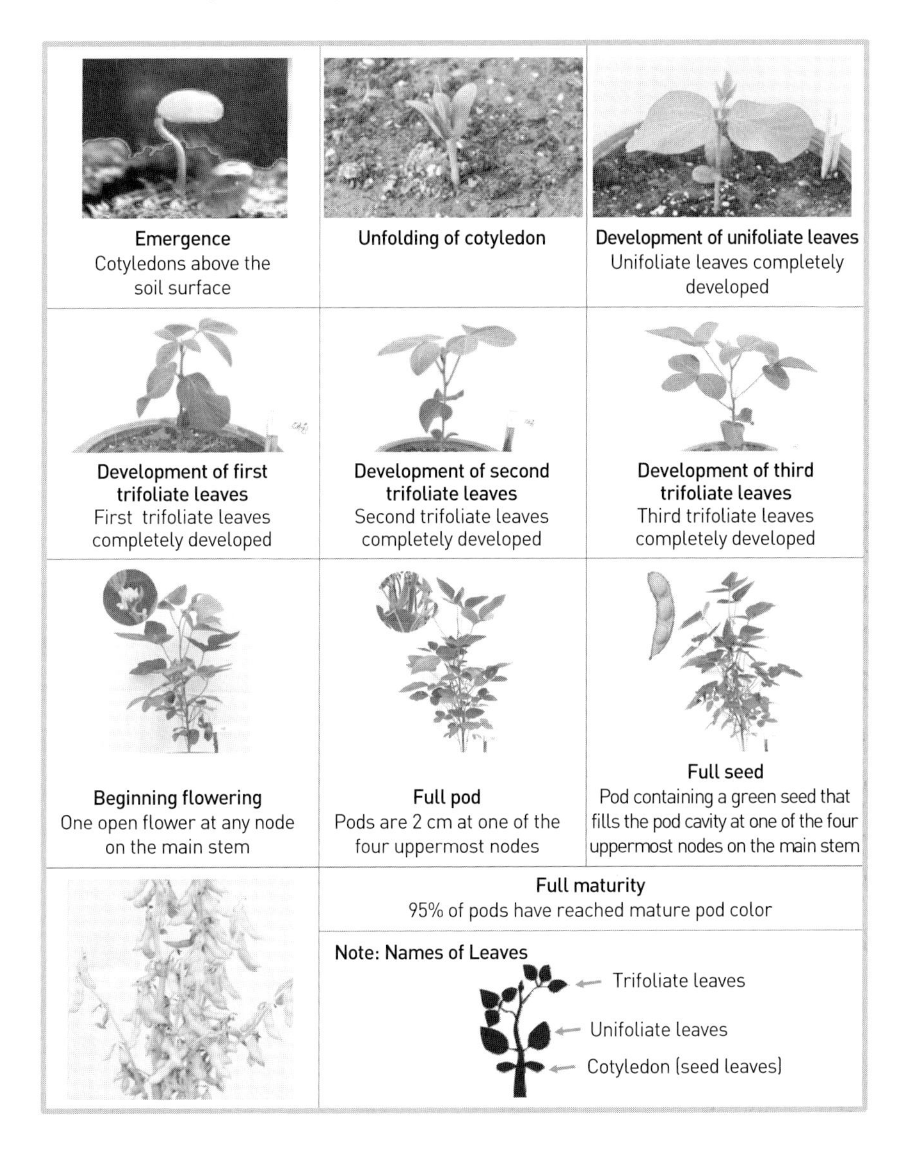

1) Han Won-yeong, Division of Functional Crops, National Institute of Crop Science (2014).

3. Variety Classification by Use

The kinds of soybeans discussed above are all soybeans bred for the making of soy sauce and other sauces that use soybean as a base. Soybeans can be divided into those used for sauces, *meju*, bean sprouts, soybean curd, and so on. They can also be divided by color: yellow soybeans (*hwangdu*), black soybeans (*heuktae*), etc. Here, "beans" refers to soybeans, which boast a protein content of 40%. Kidney beans and peas contain 60% starch and 12-16% protein, which actually makes them closer to grains.

Variety of Soybeans

- *Baektae*: Yellow soybeans and white soybeans. Widely used for *doenjang* and for sauces whose basic ingredient is *meju*; They are even called *meju kong* sometimes. *Baektae* is the variety of soybeans with the highest rate of production in Korea. They are used the most in the making of *meju*, *doenjang*, soy sauce, *gochujang*, and other sauces. It is also used to make soybean curd and is sprouted to make side dishes.
- *Seoritae*: The seedcoat of these soybeans are black and the cotyledon is green. They are also known as *sokcheong*, or "green inside." This variety has a long growing period and maturing in October, and it is harvested after the first frost. Its name, "frost soybeans," derives from the fact that it keeps maturing after a frost. When soaked in water the soybeans become quite soft and high sugar content, so this variety is usually combined with miscellaneous grains to be eaten with rice or used when making rice cake (*ddeok*). When fermented, *seoritae* make *cheonggukjang*, a fast-fermented paste that is not only good for one's health but also is more delicious than making the same paste from *meju*. The high levels of anthocyanin create pigmentation that led to the nickname "black soybeans," and the repeated and consistent usage of this anthocyanin-rich variety over long periods of time has a preventative effect on aging. *Seoritae* are rich in protein and vegetable fats and contain vitamins necessary for metabolism, especially vitamins B1, B2, and niacin.

- *Seomoktae*: This variety is also called "rat's eye soybeans" because the soybeans are small and black. The people have often called this the "medicinal soybean." it is well-known that if one applies sulfur on the soil when cultivating these soybeans, their medicinal properties improve. The taste of these soybeans is sweet, and they contain no toxins. These soybeans are related to the kidneys, such that eating them helps ease kidney disease, and the minerals in the soybeans help the body rid itself of toxins as well as improve blood circulation. If fried, they warm the body, and if taken mixed with alcohol, they have a positive effect on the nervous system. Boiled or steamed *seomoktae* mixed with ginger and salt to make medicine has a very cooling effect on the mind. Making porridge from these soybeans relieves diseases symptomized by thirst, and making sauce with them brings equilibrium to one's disposition. "rat's eye soybeans" have 5-6 times the amount of isoflavones as other soybeans.

Types of Sauces

Fermented soybean sauces are among the most representative traditional foods of Korea. Even now, the basis for breeding soybeans is for the use of sauce-making.

Characteristics of Soybeans for Use in Sauces

Soybeans for sauces must have high protein content and large seeds[1)] of yellow seed coat and hilum in order to be treated as a high-quality product. Aside from these characteristics, others to take into consideration include taste, smell, color, yield value, and, uniquely to soybeans used for sauces, not only large seed size, but also high absorption rate, softness, thinness of seed coat, and suitability for processing.

---

1) One hundred seed weight is the method of measuring the weight of seeds. Per every 100 seeds, if the weight is 40 g or more, the seeds are considered extra-large; if 25-40 g, the seeds are large; if 15-25 g, the seeds are middle; if 10-15 g, the seeds are small; and if 10 g or less, the seeds are smallest.

Good Varieties for Sauces

In pursuit of ingredient characteristics that might improve the taste of *meju*, 21 varieties of soybeans from domestic and abroad were investigated for over 30 qualities apiece at the Crops Testing Platform in 1998. The first selection was also chosen for suitability for manufacture of high-quality *meju*. The taste-related ingredients of *meju*, such as free amino acids and formol type nitrogen, were also investigated for potential aspects of improvement based on length of fermentation, and the varieties chosen as definitively excellent for producing *meju* turned out to defy expectations: they were *Danbaekkong* (14.9 g/100 seeds), *Duyukong* (20.5 g/100 seeds), and *Danyeobkong* (15.3 g/100 seeds), all belonging to the middle seed varieties. In the future, soybean varieties used for sauces will be selected not for their external qualities, but for how well they are able to be processed into *meju*, *cheonggukjang*, and other sauces. However, because it is difficult to distinguish between imported and domestic soybean varieties of middle and small size that are good for manufacturing, marketability is low. One way of solving problems like this will be to create cultivation fields and then process and ship the products from here.

Soybean curd and Soymilk

The quality of soybeans used as raw materials in the processing of soybean curd, as well as the processing itself, influence qualities of the resulting soybean curd, such as yield, texture, flavor, and shape. Among the elements that make up the chemical composition of soybeans, lipids and protein remain in soybean curd and soymilk. Among carbohydrates, fiber remains in soymilk and the residues, whereas sugars are compressed and mostly expelled in the coagulation process.

Soybeans Used for Soybean Curd and Soymilk

It is good that soybean curd's raw materials are high in proteins and lipids, and the higher the water-soluble protein content, the higher the yield of curd. The carbohydrate content of raffinose and stachyose is low, and soybean curd's nutritional qualities can be improved when made with soybeans that have the nutritional advantages of low phytic acid, which prevents the absorption of necessary minerals, and low trypsin inhibitor, which prevents good digestion. In addition, soybean varieties that have a comparatively rapid absorption rate during soybean curd production when the soybeans are dipped into water and then ground provide for easy soybean curd manufacture. When the calcium and crude fiber content of the seed coat is high, the absorption rate decreases, and also the thickness of the seed coat and the rate at which it sheds its seed coat have an influence on soybean curd yield.

Excellent Varieties for Soybean Curd and Soymilk

When considering protein content and bean curd yield, *Baegunkong*, *Danbaekkong*, and *Jangyeobkong* are excellent soybean varieties. *Danbaekkong*, which was released in 1993, has a protein content of 45% or more, 13% higher than that of standard soybeans, but its flaw is that it is a small soybean, only 14 g by 100-seed weight. From the late 1980s in order to foster varieties without a beany flavor, lipoxygenase was rendered inert in the *Jinpumkong* and *Jinpumkong* 2 varieties, which were then propagated. An advantage when processing these two varieties into soybean curd was the ability to drop the heat treatment that previously was necessary for removal of the beany flavor, but these varieties were not useful in the manufacture of soymilk because heat treatment is necessary for processing soymilk regardless.

Soybean Sprouts

Before the use of plastic film houses to grow vegetables, soybean sprouts, along with *kimchi*, were the most important source of vitamin C available from winter vegetables.

Characteristics of Soybeans used for Sprouts

Because the hypocotyl (the stem below the seed coat and above the roots) is the edible part of the plant used for bean sprouts, soybean varieties used for sprouts must be able to produce hypocotyl very quickly. Because they should be able to germinate the raw materials rapidly, small soybeans have markedly higher production rates than large soybeans when it comes to sprouts. However, when eating soybean sprouts it is good to have a relatively low amount of fiber, since too much fiber makes them tough, but on the other hand, a certain amount of fiber is necessary to prevent the sprouts from breaking. When choosing soybeans for sprouts, plant characteristics also include preference for color and gloss, such as a silver stem with golden cotyledons.

Good Soybean Varieties for Sprouts

Development of good soybean varieties for sprouts began from the 1970s, but until the early 1980s the aptitude for cross bred varieties was low, and land races were used as the raw material. In 1984 the first *Pangsakong* soybean was made by a mutation breeding method, and in 1986 due to the dissemination of the *Eunhakong* variety, which flowers early, land races were replaced by *Eunhakong* and cultivation of this variety began increasing on farms. In the early 1990s more than 50% of farms cultivated land races such as *Junjeori* and *Orialtae* because the promoted varieties, when compared to the land races, had large seeds and yield and quality were not as high as that of land races. However, in the late 1990s excellent varieties for sprouts emerged, such as *Pungsannamulkong*, *Somyeongkong*, *Sowonkong*, and *Doremikong*, and the 2000s brought *Sinhwa*, *Pungwon*, *Haepum*, *Janggi*, and others, and now most farms cultivate these varieties. In particular, breeders looking for large-scale contracts prefer these varieties for their high level of purity, strength in production volume, and resistance to pests and diseases.

Boiling with Rice

Colored soybeans, including black soybeans, have been used in soybeans-with-grains production in Korea for cooking together with rice, barley, and miscellaneous grains as a staple food, as well as for a traditional black soybean and soy sauce side dish, for mixing into rice cake, and for use in pastries and medicines.

Characteristics of Soybeans used for Boiling with Rice

The colored soybeans used for boiling with rice include those with green cotyledon, which are popular at the market because the seeds of these plants are generally large and soften well when cooked with rice. There are various elements that affect seeds in multiple ways in terms of cooking and eating qualities, including seed shape, flavor, and type of tissue. Accordingly, goals for breeding of colored soybeans entail large seeds, high water solubility, good softening capability, and pigmentation.

Good varieties for Boiling with Rice

Currently there are 14 varieties being released, including *Geomjeongkong* 1, *Geomjeongkong* 2, *Ilpumgeomjeongkong*, and black seed doat with green cotyledons *Cheongjakong*, *Heukcheongkong*, *Cheongjakong* 2, and *Josaengseori* varieties. The development of these varieties has achieved considerable achievements, but generally they are susceptible to lodging and their aptitude for use in cooking with rice is a bit lower than that of land races, and some also have the problem of maturing too late.

Vegetable and Early Soybeans

Vegetable soybean have been used from early days in Korea for cooking with rice or folding into rice cake, but it is not known for certain when they began to be used. Vegetable soybeans began to be produced for commerce in the 1990s, and the number of farms bringing these beans to market exploded. Now it is estimated that hundreds of hectares are being cultivated, though there are no definitive statistics at this time.

Characteristics of Vegetable Soybeans

Vegetable soybeans have a distinctive flavor and are full of protein, fat, and vitamins A, C, and E. Characteristics that make vegetable soybeans high-quality include 1) Soybean pods are large, 2) Soybean pods are decidedly green, but the seed coats can be yellow or green, 3) the pubescences on the pod are gray, 4) the ratio of seeds to pod are 2- 3:1, 5) the carbohydrate level of the seed is high, affording a sweet taste, 6) amino acid content is high so that glutamic acid and others contribute to good flavor, 7) the taste is good, and 8) texture is good.

Good Varieties for Vegetable Soybeans

Among the varieties bred in and after the 1990s were *Hwaeom-putkong* and *Seokryangputkong*, native pure-line isolated *Hwaseong-putkong*, and cross bred *Sillok*, *Seonnok*, *Danmi*, and *Dajin*. Recently in the south, after the increase in cultivation of onions and garlic, early soybeans are being planted as a follow-on crop, but there are

Galmikong, Gapsanjaerae, Geomjeongkong 1, Geomjeongkong 2, Geomjeongkong 3, Geomjeongkong 4, Gyeongdu, Kwangkyo, Gwangdu, Kwangankong, Geumgangdaerip, Geumgangsorip/ Geumgangkong, Geumdu, Geumsanjong, Namcheonkong, Namhaekong, Dagi, Daol, Dajangkong, Dajin, Dachae, Dankyeongkong, Danyeobkong/ Danweonkong, Daemang, Daepung, Daehwangkong, Togyukong, Doremikong, Duyoukong, Mallikong, Milyangkong, Pangsakong, Baekbamkong, Baegcheon/ Beomkong, Pokwangkong, Boseok, Bukwangkong, Buseok, Samnamkong, Sangdu, Saebeolkong, Saealkong, Seonam, Seonnok/ Sodamkong, Sorok, Somyeongkong, Sowonkong, Sojin, Soho, Singi, Anpyeong, Yuwoldu, Yuku 3 (Rikuu 3), Eunhakong, Iksan/ Ilmikong, Jangdanbaekmok, Jangmikong, Jangbaegkong, Jangwon, Geomeunkong, Bulkong, Sokpureunkong, Yuwolkong, Huinkong, Jinmi/ Jinyulkong, Jinpumkong, Jinpumkong 2, Cheongdu 1, Chungbuktae, Taekwangkong, Paldalkong, Hannamkong, Horang-ikong, Hojang, Hwangkeumkong

Representative Soybeans Preserved at the Agricultural Genetic Resources Center only eight varieties: *Keunolkong*, *Saeolkong*, *Daol*, *Geomjeongsaeol*, *Hanol*, *Hwanggeumol*, and *Chamol*. On the other hand, *Keunolkong* and other cultivated early soybeans can also be used in place of vegetable soybean varieties. Vegetable soybean and early soybean varieties are ecologically nearly the same, both being early maturing varieties, but their usage differentiates them. Currently the most pressing issue for early soybeans is developing varietals that succeed at preventing disease in the seeds.

Land Races Preserved at the Agricultural Genetic Resources Center

Korea has a greater variety of soybean genetic resources than any other country in the world, and currently the Agricultural Genetic resources Center has 8,132 land races stored. Unusual native soybean genetic resources were used as raw materials for traditional foods as well as home remedies, and thus consumer preference for these varieties of soybeans has been high. Restoring native soybean varieties, and with that boosting regional specialty products, are important works to foster.

Pedigrees of Soybean Cultivars

Pedigrees of soybean cultivars records for soybeans are similar to the genealogical records many Koreans have in their homes. Looking at these records for soybeans one can find the ancestors of individual variety and estimate the hereditary distances between them and other varieties. The pedigree is a reference that breeders can use to support genetic varieties by checking whether varieties descend from a common ancestor, in which case cross-breeding should be avoided.[1)]

1) Kim Seok-dong and Lee Yeong-ho, Breeding and cultivation of soybeans, *Kong* (Soybeans), Korea University Press, Seoul, 137-216(2005)

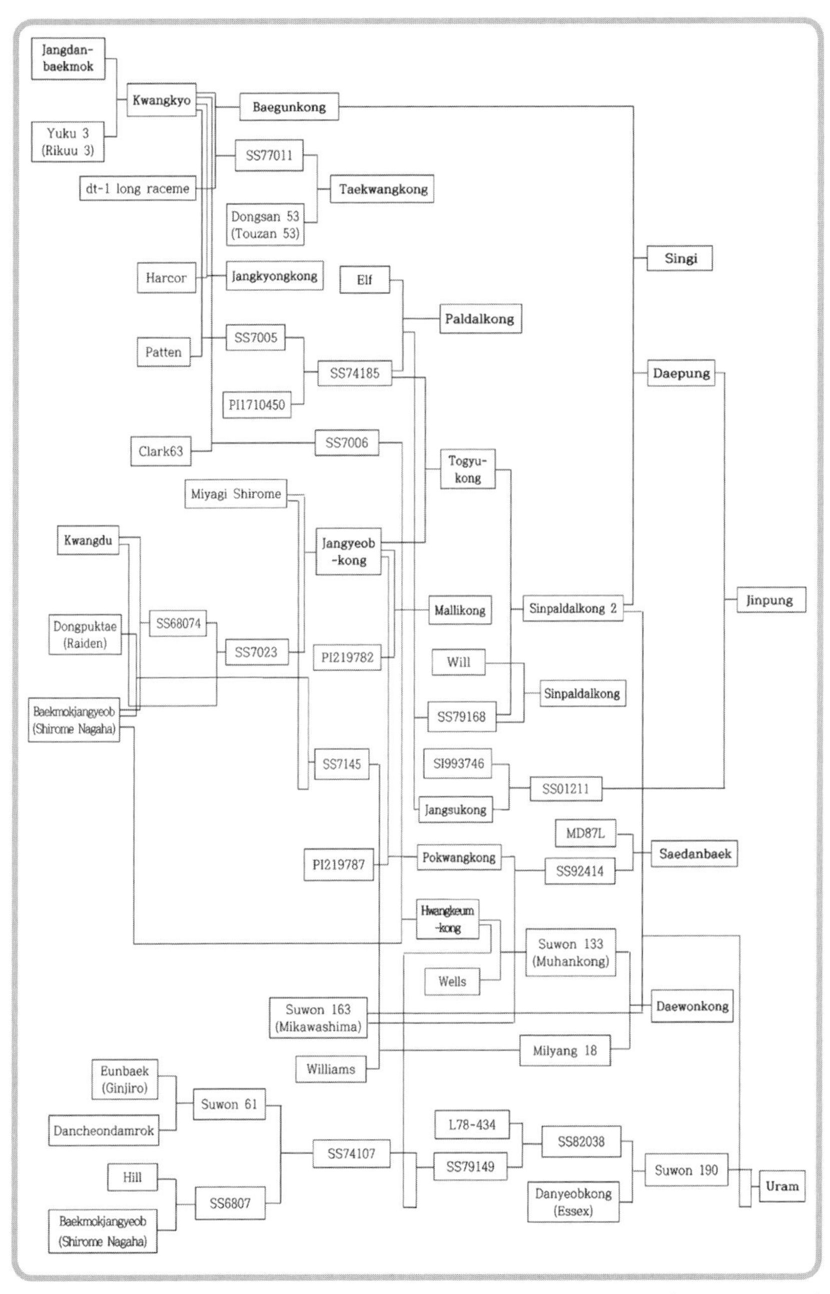

Pedigrees of soybean varieties for soybean sauce and soybean curd (Lee Yeong-Ho).

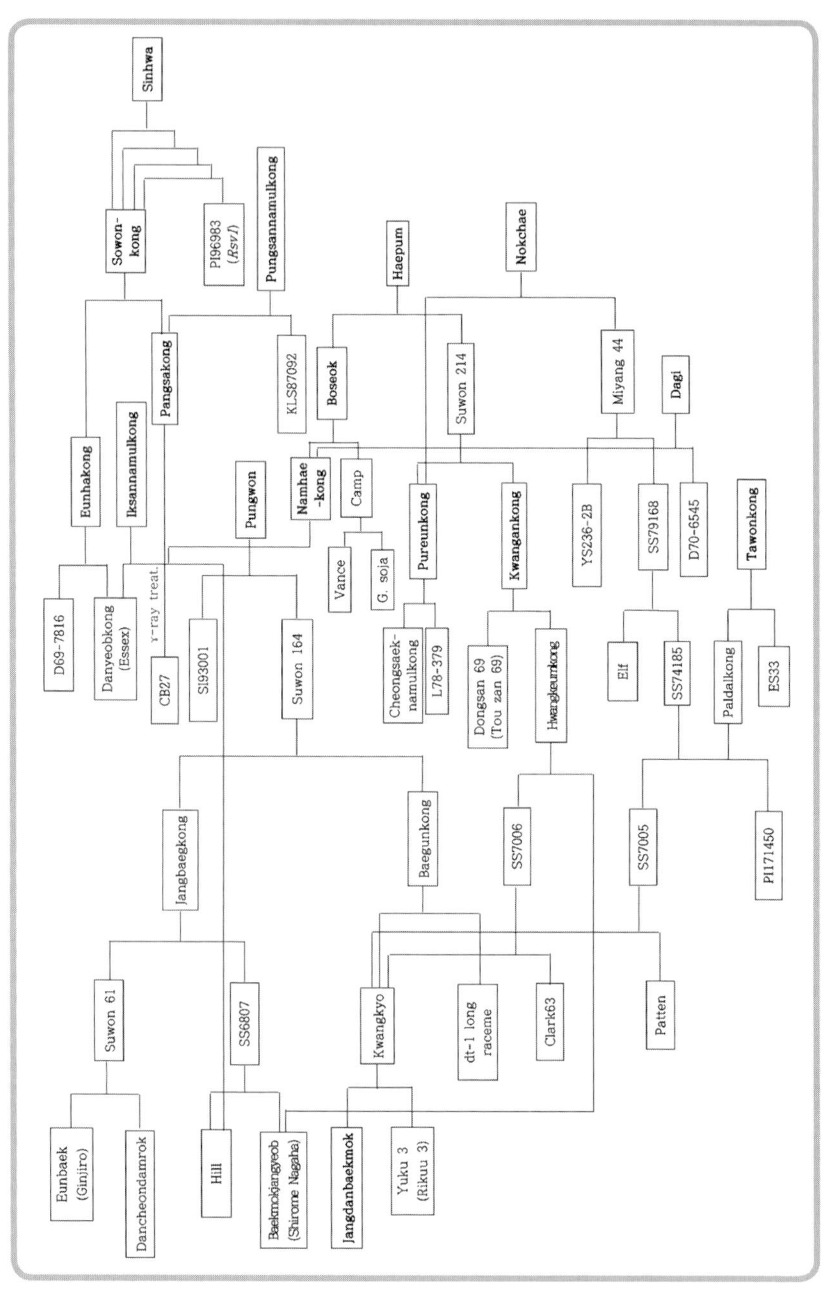

Pedigrees of soybean varieties for bean sprout (Lee Yeong-Ho).

Korean Native Soybeans

As the place where soybeans originated, the uses of soybeans in Korea are quite varied, as are the varieties and their names. The names mostly derive from regional areas, patterns, and colors, as well as their use and perhaps their sowing time, maturation time, and pod setting form. Soybeans have been grown in the north central area of the country and along mono-crop belts in the mountains. In the south varieties with small seeds have often been cultivated.

Soybean Varieties by Color and Pattern

Saealkong, seonbibamkong, cheongtae, ajuggarikong (ajuggaribamkong), heuinkong, nureonkong, geomjeongkong (heuktae), parangkong, saeparangkong, geomeunbamkong, bamkong, jajukong, sokpureunkong (sokcheong), bichukong (bichwikong), pureudaekong, subaktae, jagalkong, daechukong, daechubulkong, jwinunikong (seomoktae), aljongdarikong, nunggamegikong (nunggamjangikong).

Soybean Varieties by Harvest Season

Oltae, 40ilkong, swinnalgeorikong (50ilkong), yuwoldu (yuwolkong), seorikong (seoritae).

Soybean Varieties by Cooking Use

Ddakkong, gomulkong, parangomulkong, namulkong (jiremikong, jilgeumkong, jireumkong), mejukong, yakkong, bammitkong (baemikong, bapkong, banmikong).

Soybean Varieties by Collection Place

Gapsantae, Cheongsantae, Jeongseonkong, Buseok, Jangdanbaekmok, Yeongyang, Ulsan, Haman, Iksan, Dancheon.

Soybean Varieties by Type of Seed and Plant

Bogakdarikong, junjeorikong, jomkong, napjakkong (napddegikong, napdeurekong, napjjorigikong, napjireugikong), hanagarikong, buchaekong (or maendeuramikong).

Soybean Varieties by Planting Place

Boribatkong (boriganjagyong), nondureongkong (dureongkong).

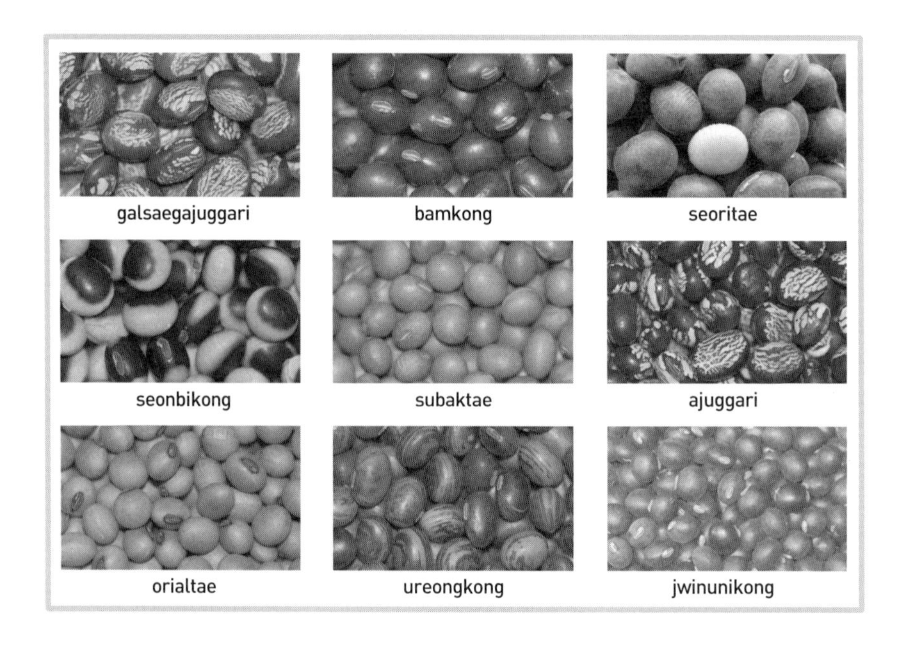

Yeongju Buseoktae 1

Buseoktae soybeans cultivated in the Yeongju region were not produced through cross-breeding, but rather via the pure-line selection method. Standard soybeans weigh 20 g per 100 seeds, while *Buseoktae* 1 weighs about 40 g and boasts excellent flavor and nutrition. Yeongju has long been a chief producing district that has kept the production of soybeans alive. The *Buseok* variety was collected in Yeongju, North Gyeongbuk

Province in 1948 was by 1960 promoted throughout the North Gyeongbuk area and propagated across farms there. The efforts of Yeongju city brought about an opportunity to establish the Soyworld Science Museum there, with a goal to get a foothold on soybean culture, tourism, education, and business in the region through joint research and mutual cooperation towards fostering international research and creating a business mecca for soybeans. In order to overcome the hereditarily unstable weak point of *Buseok* soybeans, in April 2009 the city of Yeongju signed a memorandum of understanding with the National Institute of Crop Science, Rural Development Administration, which has the best breeding techniques in the nation. As a result, *Buseok* soybeans, which had contained various mixed genes, were improved through pure-line selection. In order to apply for *Buseoktae* 1 as a new variety, breeding techniques had to be supported, and this was done in such a way that it came to be grown according to specifically-tailored cultivation techniques.

4. Farming and Cultivation Methods

The extent of soybean cultivation in Korea covers about 10% of the total upland of 755,000 hectares. Among fields planted with soybeans, most use the mono-cropping system, and location conditions are very poor. Looking at the distribution of location of these fields, 69% exist on farms in mid altitude or in the mountains, and it is not easy to engage in mechanization on such farms, watering poses a problem, and it is easy to incur damages from conditions of frequent drought. In order to stabilize the supply and demand of rice, the Ministry of Agriculture, Food, and Rural Affairs encourages farms to plant corn or soybeans rather than rice–decreasing rice production and increasing the production of other food crops–in order to maintain a high level of overall food self-sufficiency.

Soybean Farming

The question of when is the best time to plant soybeans depends on each variety's maturation time (early, middle, or late), meteorological conditions, cultivation method, cropping system, and more. Generally speaking, soybean planting is possible when the ground reaches a temperature of 15℃ or higher. The most important element in achieving a high yield of soybeans is securing full germination through the uniform emergence of sprouts. When soybeans are planted, 2-3 seeds are planted in each hole, taking into consideration failure to emergence and damage that may be incurred by harmful insects. The soybean planting season in Korea lasts for about three months, from the beginning of April to the beginning of July, and there are significant differences in yield depending on when seeds are planted. This long planting season is one of the important characteristics of soybeans. Knowledge of this characteristic can help increase arable land by employing crop rotation with other crops. One cultivation method that has generally been adopted in most areas around Korea is to plant soybeans after barley. There are various other methods of cultivation according to different regions in the country, such as crop mixing with corn and soybeans, intercropping soybeans just before barley harvest, or crop rotation with rice, barley, and soybeans.[1)]

Variety Selection

Soybeans are bred according to their usage. Varieties are determined based on their destined usage as *doenjang*, soybean curd,

1) Hong Eun-hee, The history of soybean cultivation, *Kong* (Soybeans), Korea University Press, Seoul, 103-136(2005)

bean sprouts, vegetable soybeans, etc., and for whether their cultivation method is mono-cropping, double-cropping, etc., and how quickly they mature in their given form of cultivation.

Planting Time and Methods

The optimal planting time for soybeans in the middle of the country is early June; in the south it is mid June. In cases where soybeans are planted as a follow-on crop to spring potatoes, spring corn, onions, or garlic, the planting time moves to late June through early July, but certain varieties, where possible, can be pushed back further still, to be planted in mid July, in which case fall onion and garlic planting may be considered after the harvest of the soybeans. Soybeans should be planted 3-5 cm deep. If planted too deep, emergence will be unfavorable, and if planted too shallow, the moisture in the soil will be insufficient and germination will be unfavorable. Planting tools are widely used, either by attaching them to a tractor or using them by hand.

Plowing and Fertilization

Deep tillage improves the porosity, hardness, water permeability, and other aspects of the organic matter content of soil. Tillage also promotes the vigorous health of the plant by expanding its rooting zone, so plowing in the fall or early spring is good for soybeans, especially by increasing the permeability of paddy soil and thus encouraging proliferation of root nodule bacteria, which provide soybeans with the 50% supply of nitrogen needed for their lifespan. Tillage can also have a drying effect on wet paddy soil. Most paddies have clay soil, so that when it rains, the stratum of arable soil hardens, thus making planting difficult and soil preparation necessary. It has been assumed that soybeans can grow in poor soil, so they have in the past been cultivated with no soil improvements, but the amount of nitrogen absorbed by soybeans is 50% fixed: 40% from the fertility of the soil, and 10% from fertilizer. If soil fertility is to improve, then soil amendments must be made. The standard fertilizer for soybeans, since each segment of soil has a different level of fertility, is controlled according to the fertility of a given part of the soil; most farms today use complex fertilizer with a standard amount of required nitrogen and phosphorus and potash as a supplement with single fertilizer to prevent a nitrogen overdose, which could lead to overgrowth or collapse in the face of too much rain or wind.

Soybean Harvest and Management

After harvest, soybean water content stands at about 17%, so it is good for sorting and storage to dry the soybeans. Early on, rather than using mechanical drying machines, soybeans would be left in the fields to dry out naturally up to 80%, and the remaining 20% would be dried using a drying tool or in the shade so that the soybeans could be sorted and stored well. Threshed soybeans to be used as seeds are dried to less than 13% moisture content, while care is taken not to mix in other varieties or soybeans damaged by insects. Allowing soybeans to dry in the sun supports good germination, and in areas where a drying machine is necessary due to weather conditions, the temperature should be 30-40℃ and air-dried. If dried too fast and at too high a temperature, cracks will appear in the pods and the quality of the seeds will decline.

Sorting Soybeans

After harvesting and drying are complete, the next step is sorting. Sorting is done either by machine or by hand, and in order to prevent seed contamination or mixing, sorting must be supervised. Machines sorting soybeans to be used as seeds from those slated to go to market must be thoroughly cleaned afterwards. The surrounding area must be closely observed so as to not mix soybean varieties or allow foreign substances to get mixed in with the soybeans. Finally, the soybeans must be sorted by size in order to command the best price at market. Most farms that cultivate soybeans to be processed into soybean curd, sauces, and sprouts employ a sorting machine in order to standardize the quality of their product, while farms that produce soybeans for boiling with rice must sort the soybeans by hand due to the necessary relationship between shape and volume. It seems that in order to improve the quality of soybeans, the use of machines for thorough quality control will become a necessity.

Storage of Soybeans

After harvesting soybeans, storage plays an important role in preserving the quality of the product. The season for storing soybeans on the farms that produced them is winter (December-February). Farms that cultivate soybeans but do not have low-temperature storage facilities store their product in open warehouses instead. When the air temperature rises, the quality of the stored soybeans begins to decline: warmth begins a metabolic process that consumes the soybeans' protein and decreases their germinative energy and nutritional profile. When storing soybeans for seed, the life expectancy of the seeds largely depends on storage temperature, humidity, and seed moisture content; if the temperature and humidity are high, life expectancy will be reduced, so well-dried seeds must be stored in dry and cool conditions. In order to prevent harmful pests during long-term storage, it is helpful to fumigate the seeds for 4-7 days. Seeds stored for a year or longer must be kept at a temperature of 5℃ or lower, with a relative humidity of 45-50%.

 *Ssiojaengi* (small straw bags) Stored soybean seeds.	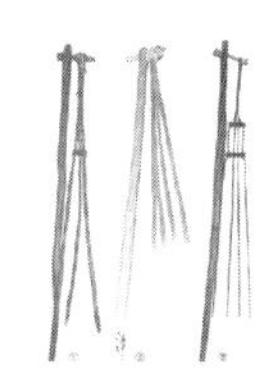 *Doriggae* (flails) Threshed the grains from the plant.	 *Punggu* (winnowing machine) Sorted grains from dust, chaff, and hulls. Used for rice, barley, soybeans, adzuki beans, millet and other grains.
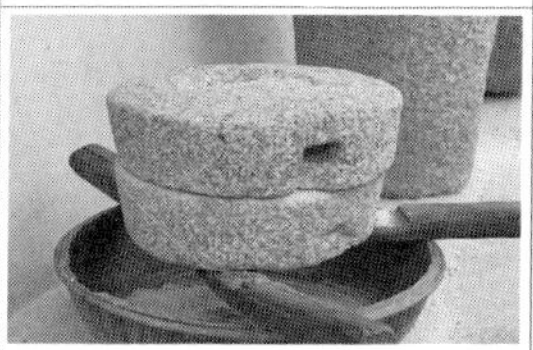 *Maettol* (millstones) Ground grains into flour.	 *Kongnamul siru* (bean sprouting jar) Earthenware jar used to sprout soybeans.	 *Dubuteul* (soybean curd mold) Mold used to shape soybean curd during the congealing process.
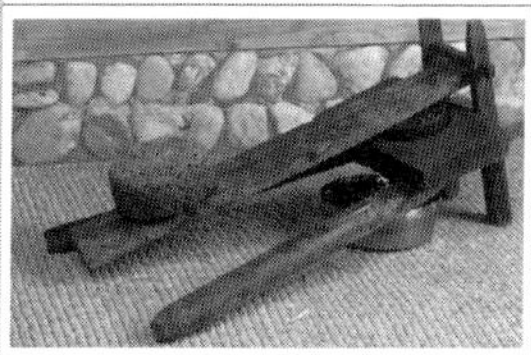 *Gireumteul* (oil press) Pressed seeds such as soybeans, sesame and perilla to make oil.	 *Jangdok* (crock) Earthenware jar used for making or storing sauces such as soy sauce, doenjang (fermented soybean paste), and gochujang (red pepper paste).	 *Dukbaegi* (earthenware pot) Used for boiling stews or making jorim (food boiled down in soy or other sauces).
		*Jeolgu* (large mortar and pestle) Used when pounding soybeans for meju (fermented soybean cakes). Mortar could be wood or stone.

Tools of Soybean Culture

Chapter 04

Processing & Utilization of Soybeans

1. Utilizing Soybeans

Soybeans have long been consumed in Northeast Asian regions, which boast the most advanced and varied form of soy utilization. Their actual methods of utilization may greatly differ from region to region, depending on their historical and cultural background. Soybean processing can be largely divided into two categories. One is cooking it without changing its original form as seen in *kongbap*, cooked rice mixed with soybeans, and *kongjaban*, cooked soybeans seasoned with soy sauce. The other it by changing its original shape, i.e., grinding, fermenting, dissolving, sprouting, extracting, separating to dry, solidifying, preserving with salt, or texturizing. Boiled soybeans, *doenjang* (fermented soybean paste), soybean flour made by grinding peeled soybeans, and whole soybean curd are examples of products utilizing the whole components of soybeans. Bean sprouts are an example of using a sprouting method, soybean curd makes use of a solidifying process, soybean oil and soymilk are made by an extraction method, soybean flour is made by grinding, and powdered products such as condensed or isolated soybean protein products are made by separating and drying method.[1)]

<u>**Products Utilizing Key Soybean Components**</u>

Soybeans in *kongbap*, *kongjaban*, *doenjang*, soymilk, bean-curd dregs, and full fat soybean powder made by grinding peeled soybean are examples of the products intended for taking in the whole components of the bean, while soy sauce, *dubu*, *yuba* (*dubu* skin), and texturized, defatted soy flour utilize protein. Examples of products using fats include soybean oil, soybean butter, tocopherol, lecithin, and fatty acids, while isoflavones, pinitols, oligosaccharides, and peptides are made by the bean's functional components.

2. The Soybean's Great Transformation

Processing the soybean is indispensable because soybeans are rarely digestible if eaten raw. *Kongbap* is 65% digestible, and *cheonggukjang* (fast-fermented soybean paste) and *dubu* are 85% and 95% digestible, respectively. Protein contains no component producing a flavor in itself. However, if it is broken down into its separate elements, i.e., peptides and amino acids, it produces a savory taste. It is the role of traditional soybean *meju* to induce the proliferation of various microorganisms actively multiplying in the bean by creating a natural environment that attracts them, so that soybean protein can be broken down into its separate elements with the help of enzymes in these microorganisms. For this reason,

---

1) Kim Woo-jeong, The processing characteristics of soybeans, *Kong* (Soybeans), Korea University Press, Seoul, 217-312(2005)

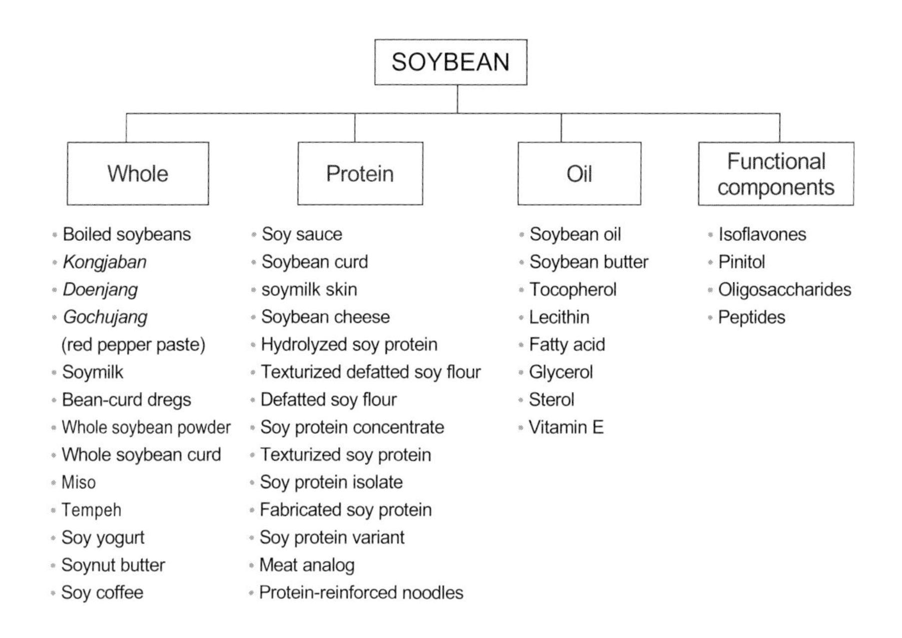

the quality of fermented soybean food is closely related to the fermentation of soybean *meju*. The more protein fermented foods such as *doenjang* and soy sauce contain, the better their quality becomes; and the fat content and a composite of fatty acids are essential to the manufacture of soybean oil.

The hydration speed, or absorption of water when soaked, the cooking speed, and the cooked soybean's texture play important roles in determining the quality of soybeans when cooked with rice. However, the quantity of water-soluble protein content and solids content are equally important in the case of soymilk or *dubu*.

Also, the elimination of anti-nutritional factors, such as trypsin inhibitors and phytic acid, and the effect of functional components in preventing chronic diseases are recognized as major determining factors of quality.

Processed Soybean Foods

Manufacturing Soymilk

Soymilk is produced by grinding soybeans in water and then heating the resulting mash to squeeze out liquid from dregs (residues). Soymilk is a quality food full of healthful dietary protein and other nutrients. In summer, thoroughly washed soybeans are soaked for 7 to 8 hours (24-hour soaking is necessary in winter season), grind in a hand mill and then boiled. Deodorization of beany flavor can be done through such boiling, and much of the protein is dissolved out from the dregs in the process. After boiling, the mash is put into a pouch of cotten cloth to squeeze the soy milk from it. It is easier to squeeze soy milk while the pouch containing the boiled mash is still hot. Soymilk is used for nutritious beverage, and the residue(*biji*) for Ijigae and side dishes.

Manufacturing Soybean curd

Soak the soybeans in water sufficiently to grind and then boil them. The boiled soybeans should be filtered to remove bean-curd dregs, then add a coagulant to the resulting soymilk, and once it has set, press it into a rectangle. Insoluble protein, poly-saccharides, and considerable amounts of fat are removed along with the bean-curd dregs, and the remaining fat and carhohydrat are left in the curd as the water-soluble protein begins to set. The soybean manufactured and distributed in Korea includes unpressed *dubu*, regular *dubu*, and silken *dubu*. Regular bean curd, which is most commonly consumed, has a light taste and is relatively solid and slightly elastic, with a rough texture. Meanwhile, silken *dubu* has a uniform texture, and its surface is smooth and soft. Unpressed *du-*

bu is less tender than silken *dubu*, but much less solid than regular *dubu*. In the past, only roughly texturized, unpressed *dubu* was made, but unpressed *dubu* with a uniform texture has been manufactured recently. Processed soybean curd products include *dubu* skin, *saengyang* (made by frying only the skin of *dubu*), deep-fried *dubu*, freeze-dried *dubu*, egg silken *dubu*, powdered *dubu*, bean curd noodles, fortified *dubu* (soybean curd fortified with vitamins), fish *dubu*, fermented *dubu*, *sufu* (*dubu* fermented for extended shelf life), milk *dubu*, mixed *dubu* (cooked and condensed soymilk mixed with vegetables and seaweed), and flavored *dubu*.

Manufacturing Soybean Sprouts

Soybean sprouts are the most traditional primary processed food, and nutritive components vary drastically during the soybean's sprouting process. Fat contained in the seed decreases while dietary fiber increases greatly. In particular, vitamins C and B_2 and asparaginic acid are rapidly formed during the sprouting process, increasing the plant's usefulness as a vegetable. So too are carnitines, which are good for fatigue recovery. Asparaginic acid, which is known as a hangover cure because it improves the function of an alcohol-degrading enzyme, is contained plentifully in the fine roots of bean sprouts. Potassium, which lowers blood pressure, and isoflavones, which have carcinostatic effects, increase during the water-soaking and sprouting processes. The content of vitamin C in soybean sprouts reaches its highest level on the 7th day of sprouting. Meanwhile, the longer the sprouting period, the higher the level of asparaginic acid becomes.

Fermented Food

Manufacturing *Doenjang*

Meju that have been used to produce soy sauce in a crock are taken out and crushed, and then salt and soy sauce are added and aged in crock to produce traditional, home-made *doenjang*, as mentioned in chaper 2. However, factory-made *doenjang* are manufactured using *koji* (Using a*spergillus oryzae)* in place of the *meju.* Put *koji*, boiled soybeans and brine into a container and press it down tightly with salt spread over it, then cover with a plastic sheet and place a stone weight over it. Ripening for about 2 months at 25 to 30℃ causes the soybean malt layer to form amino acids through the action of proteinases, producing a savory taste. Ripening takes place through the interaction between the lactic acid produced by halotolerant lactic acid bacteria and alcohol produced by halotolerant yeast. Occasional mixing is done when the need arises, and then it is ground and packed, turning it into an improved *doenjang* product. Improved *doenjang* products have been made using the same method as miso (Japanese bean paste) since the introduction of *koji.*

Manufacturing Soy Sauce

The major difference between factory-made and traditionally-made soy sauces lies in the wheat-based manufacturing process of *koji*. For raw, genuine soy sauce, cooked defatted soybeans are mixed with the same amount of roasted wheat along with *koji* to ferment and ripen for 6 months. Food additives, including saccharides, are added to this raw soy sauce and then sterilized and filtered to produce soy sauce. Amino-acid soy sauce is produced by forming amino acids through the hydrolysis of defatted soybean or gluten, a byproduct or wheat starch by using hydrochloric acid, which, in turn, have the liquid separated out after using a neutralizing agent. *Koji* is made using defatted soybeans and wheat to turn liquid into a soy sauce mash, after which enzyme preparations are added to produce enzyme-decomposition soy sauce. Mixed soy sauce is a product made by mingling soy sauce, amino-acid soy sauce, and enzyme decomposition soy sauce at an appropriate ratio.

Types of Soy Sauce

[Categorization by Concentration]

- *Jinganjang*, or dark soy sauce: Refers to soy sauce aged 5 years or older and is sweet in taste but dark in color. It is mainly used for preparing *yaksik* or *jeonbokcho*.
- *Joongganjang*, or medium colored soy sauce: Refers to soy sauce aged 3 to 4 years and is used for preparing *jjigae* (stew) or *namul* (seasoned vegetables).
- *Mulgeunganjang*, or pale colored soy sauce: Refers to soy sauce aged 1 to 2 years and is light in color and suitable for use in *guk* (soup).

[Categorization by Ingredients]

- *Joseonganjang*, or Joseon soy sauce: Only soybeans are used, with no starchy materials added. It is fermented mainly using bacteria (*Bacillus subtlis*).
- Japanese style soy sauce: Soybeans and starch materials are mixed. It is fermented using *Aspergillus oryzae*. Most of the mass-produced factory-made soy sauces are Japanese style soy sauce.
- *Eoganjang*, or fish sauce: Either the flesh and bones of fish or its intestines are used. It is decomposed and ripened with no help of microorganisms but using its own enzymes. *Eoganjang* is widely used in and around the coastal regions of China and the southern coast of Korea as well as in Japan and Southeast Asia, and the fish, its raw materials, are prepared in their entirety and sometimes without their heads and intestines.

[Categorization by Manufacturing Method]

- Traditional soy sauce: Made of soybeans only.

- Modernized soy sauce: Manufactured using whole wheat, and does not produce *doenjang*, a byproduct of soy sauce manufacturing.
- Amino acid soy sauce (chemical soy sauce): Amino acid soy sauce is also called acid decomposition soy sauce or chemical soy sauce. It is referred to as acid decomposition soy sauce in the Food Sanitation Act of Korea. Amino acid soy sauce is produced in a very short period of time using the protein raw materials, which are hydrolyzed by hydrochloric acid and then neutralized by an alkali.
- Salt-free soy sauce: This is a salt-free product intended for those patients who need to carefully control their salt intake. Such a product is produced using a method of acid decomposition, but producing it by fermentation is not possible.

Manufacturing *Gochujang* (Red Pepper Paste)

Factory-made *Gochujang* is produced by mixing cooked wheat flour *koji* with cooked wheat grains, rice, or sticky rice, and then adding salt and water until its moisture content reached to about 50%. This mixture is ground and fermented and then starch syrup and powdered red pepper are added to it, followed by sterilization at a temperature range of 60 to 70 ℃. Powdered red pepper can be added before or after ripening; either way is common practice in the manufacturing of *gochujang*.

Manufacturing *Cheonggukjang* (Fast-fermented Bean Paste)

Factory-made *cheonggukjang* is manufactured using a bacillus strain of bacteria that is separately cultivated and then added to the soybeans to speed fermentation. Traditional *cheonggukjang* makes use of the bacteria adhering to rice straw, and the modernized version of *cheonggukjang* is fermented by adding cultured bacteria. Using cultured bacteria has the advantage of producing uniform quality, thus enabling mass production.

The Science of Soybean Cooking

Soaking soybeans in water softens their texture and allows them to cook more easily, in a reduced amount of time. The soybean's speed of water absorption varies depending on its period of storage, preserved condition, water temperature, and type and quantity of soaking water. However, the absorption rate is fast for the first 5 to 7 hours at a temperature range of 19 to 24.5℃, and a saturation point is reached at about 20 hours, thereby absorbing the amount of water equivalent to or higher than 90% of the original weight of the soybeans. If heated, the proteinases in soybeans will be activated, which facilitates digestion, and the deodorization of soybeans can be done through volatilization. Swelling occurs more quickly on the surface, forming wrinkles. Controlling temperature by pouring cold water over boiled soybeans will speed up water absorption. In preparation of *kongjaban* higher concentration of sugar can also raise osmotic pressure, leading to seed cotyled one shrinkage, wrinkle formation on the seed coat, and hardening. For this reason, it is wise to pour sugar little by little and gradually increase the concentration of sugar in order to reduce wrinkle formation.

3. Soybeans, Food for the Citizens of the World

Soybeans are a wonderful source of food by themselves without any complicated processing. However, soybeans are increasingly processed in the West by separating protein and fat to manufacture bean-based processed food. 90% of the worlds production of soybean meal is being used for livestock feed, and only 0.5% of soy protein and 2.6% of soybean oil are used for food processing.

Soy foods have become a major source of protein and vegetable oil, forming a part of the normal diet in Asia, encompassing

Korea, China, Japan, Indonesia, and India; indeed, there is a variety of fermented soybean all-purpose flavor enhancer unique to Southeast Asia, which by itself is an integral part of the region's traditional food. Meanwhile, soymilk has been prepared and eaten in Korea, China, and Japan. In North Korea it is called *konguyu*, meaning milk made of soybeans. Soymilk is a water-soluble extract of soybeans and similar to cow's milk or human milk in appearance and composition. Traditional soymilk is produced by following processing steps: soaking soybeans in water, grinding, filtering, and heating them and removing the dregs (*biji*). Soybean dregs are a byproduct resulting from filtering the soymilk during the manufacturing process. They are composed of fibrous material, such as a seed coat and an embryo bud, and healthful dietary protein, fat, and inorganic substances. Soybean (residues) are used for food, but recently after having it dried and ground, it is used as a dietary fiber additive in fermented food, as a flavor enhancer, and in confectionery and bread baking.

Soybean Products in the World

China: *Doubanjiang, Sufu, Douchi*

Soybeans and soybean-based products form an important part of a normal diet in China. Chinese eat soybean soup (soymilk produced by grinding soaked soybeans) when they have breakfast at home or sometimes at a street stall or in a small restaurant on their way to work. For lunch and supper, soybean curd and *tangyup* (referring to *dubupi*, which is prepared by boiling soymilk mixed with soy flour and removing the thin layer that forms on the surface and then drying it) and *dubugan* (referring to dried bean curd prepared in a series of processing steps: wrapping the curd mixed with spices in a hemp cloth, boiling it, and then drying it) are frequently served. Meat and vegetables are often wrapped with *tangyup*, just like *kimchobap* (laver-wrapped rice with vinegar), and then fried or boiled before being served. *Dubugan* is often cut in slices and eaten fried. Chinese food based on fermented soybeans includes *doubanjiang*, *sufu* (fermented soybean curd that is often called Chinese cheese as its texture and flavor are similar to cheese) and *douchi* (豆豉). There are other fermented soybean foods in China such as *hamdusi* and *damdusi*. *Hamdusi* is the Chinese equivalent to Korea's *doenjang* and *ganjang*, and *damdusi* is equivalent to Korea's *cheonggukjang*. *Doubanjiang*, which is frequently used for spicy Sichuan food, is well known because it is served along with Mapotofu, the most famous bean curd dish in China.

Japan: Miso, Shoyu and *Natto*

Japan is a model country reputed for human longevity. The secret of longevity lies in the fact that they enjoy fermented food without strong spices, but make the most of the flavor inherent to the ingredients when cooking. *Natto* is an indispensable part of a normal diet for Japanese people who prefer cooking in this manner. Soybeans are fermented, but not to the point of becoming *doenjang*, *natto* is similar to *cheonggukjang* (fast-fermented bean paste) and maintains the original shape of the soybeans, but produces mucilage resulting from fermentation. *Natto* is most commonly eaten with cooked rice and then mixing it with a raw egg. Its efficacy is known to prevent osteoporosis and constipation, and it is also known as a diet food. Recent research favors the opinion that it improves the symptoms of hyperlipidemia if taken continuously.

Miso is a Japanese *doenjang* made by mixing boiled soybeans with rice or wheat *Koji* and salt. It is classified into rice *miso*, barley *miso* and soybean *miso*, depending on the type of *Koji*. It is largely divided into red miso and white miso according to its color. It is a fermented food commonly used in various countries in the East, including Korea, but most of the countries soy sauce is produced using a method developed in Japan ever since Japan's successful commercialization of *koji* in 1915.

Fermented and salted foods were used beginning in the Yaoi Age, by which time a typical diet consisted of staple foods and side dishes. Fermented and salted foods are largely categorized into *gokjang* (produced by fermenting rice, barley or soybeans), *yukjang* (produced using salted animal and poultry meats) and *chojang* (produced using salted fruits, vegetables, and seaweed). In later ages, miso and shoyu were developed from *gokjang*, *shiokara* and sushi were developed from *yukjang*, and *tsukemono* was developed from *chojang*. *Kinukoshitofu*, *goritofu* and *denkaku* are examples of soybean-based foods made of soybean curd. About 230 kinds soybean curd products were introduced in a book titled *Dububaekjin*, published in the second year of Tenmei (1702).

Indonesia: Tempeh and *onzm*

In Indonesia, soybeans are mainly used for soybean curd, tempeh, edible oil, fermented soybean sauce and soymilk. Soybean curd is available in every market in Indonesia, usually manufactured and distributed by a family enterprise or a small business. Unlike in Korea, very small beans are used for producing bean curd. Tempeh, a typical soybean product in Indonesia, is produced by soaking peeled beans in water and then mixing them with *ragi*[1] before fermenting for 1 or 2 days and shaping it into a cake. *Ragi* contains Rhizopus fungi, which is the fermentation fungus for tempeh. Tempeh is easily fermented at a temperature of 30℃; white tempeh is produced if white and sticky hyphae (threadlike filaments) are formed, and if fermented at a lower temperature, grey or black hyphae are formed to produce black tempeh. These hyphae stick together around the beans, enabling shape formation. Tempeh is not eaten raw; it can be eaten when fried or added to soup.

Brazil: *Feijoada*

Soybeans are well known for their anti-cancer effect. In particular, unlike yellow soybeans, black soybeans contain glycitein in their skin, which is a carcinostatic substance. *Feijoada* made of black soybeans is one of the best health foods in Brazil. This food was originally the normal diet of slaves, but has recently garnered attention as the importance of soybeans has come into focus.

Africa: *Dawadawa*

This is a traditional fermented flavor enhancer made in the savannahs of West Africa. Originally it was made using African black beans called locust beans, but soybeans are often used as a substitute raw material due to a locust bean shortage. Cooked soybeans are rolled into balls to dry in the sun. They are put into soup

1) Word for yeast. In a narrow sense it means traditional solid yeast in Indonesia, and in a broad sense it means a fermentation starter used for almost all types of fermentation.

or a sauce when cooking and eaten with the soup or sauce. These ball-shaped cooked soybeans can be made flat by hand-pressing and then dried in the sun to preserve them. Locust beans are black, so its products have a tinge of black and also give off an extremely strong smell. *Dawadawa* is an indispensable flavor enhancer that is used as a base for a sauce or stew. Fermentation fungi used for *dawadawa* are similar to those used in *cheonggukjang* in Korea, and this product emits a strong scent of rancid butter.

Nepal: *Kinema*

Kinema is similar to *cheonggukjang* and is usually eaten in winter by the Kirato tribe, who live in the central mountainous areas of Nepal. It is made by cooking soybeans, crushing them, and adding ashes for fermentation before drying them in the sun. It produces a foul smell.

Thailand: *To-anao*

To-anao refers to a rotten bean. It is made using cooked soybeans; cooked soybeans are wrapped with a banana leaves and then fermented for 3 to 4 days. Well-fermented soybeans mixed with salt and spices are pounded in mortar, and then steamed to make a jelly of red beans or dried in the sun before being eaten. It has an intense smell and is eaten by those people living in the northern mountainous areas of Thailand.

Bhutan: *Libi-iba*

Cooked soybeans are put into a round bamboo basket with a covering of cloth placed over it, and then fermented at room temperature inside a damp room. When it achieves a certain smell, it is removed from the basket to be pounded again, and then put into a jar for ripening in a warm place. Ripening takes about 1 to 3 years; the longer the ripening period is, the better the product quality becomes. Since it is stored for a long time without adding salt, it smells so terrible that even some people among those of the minority race who produce it hate the smell; it is produced only in the limited areas. The meaning of the word libi-iba is "rotten beans." It is used as a flavor enhancer.

India: *Sjache*

India is famous for its diversity in terms of race, religion, and food. A variety of beans and dairy products such as milk, butter, and yogurt are major protein sources of Indian food. *Sjache* is a food made and eaten by the people living in the Assabu region of India. It is made by fermenting cooked soybeans in a bamboo basket. Well-fermented soybeans are pounded in a mortar, and then wrapped in a round shape, using banana leaves to dry it. *Sjache* dried in this manner can be eaten and stored for a long time.[1)]

Soybeans for Industrial Use

Products Made by Isolating Soy Protein

Impurities must be removed and the beans crushed into 6 to 8 pieces before peeling. The seed coat and embryo bud are then cut into very thin slices before a solvent is used to remove fat. The bean's moisture content must be less than 12% before fat removal.

1) Cho Jung-soon, Soybean-based foods in other countries, Kong (Soybeans), Korea University Press, Seoul, 529-578(2005)

If the moisture content exceeds this limit, the beans can be dried at a temperature of 82℃ or below until they reach the desired moisture level.

- Soy Flour

Soy flour is manufactured using defatted soybean meal or normal soybeans, which are finely crushed to produce powder. It contains no gluten but plenty of protein, differentiating it from wheat flour. Also it differs from powdered skim milk which contains cellulose. Since soy flour is made by grinding peeled soybeans, its components are similar to the composition of the bean's cotyledon. Defatted soy flour contains almost no fat but plenty of protein due to the removal of fat. Major quality control regulations stipulate that the fat content of defatted soy flour should be 1% or less, while soy flour made only by grinding peeled soybeans contains at least 18% fat.

- Soy Protein Concentrate

Soy protein concentrate consists of protein and insoluble carbohydrates, and was first commercialized in 1959 in the United States. Soy protein concentrate made in accordance with relevant quality standards has a protein content of more than 70% in a dry base because most of the fat and water-soluble non-protein substances are already removed. Soy protein concentrate has a nitrogen solubility of more than 65% and good emulsifying and water-binding capacity, and is therefore added to food such as sausages. It is also used in products requiring casein or skim milk such as meat products, nutrient-fortified drinks, and soup bases.

- Soy Protein Isolate

Soy protein isolate was originally called soy casein in 1903, and it was first used in binding agent with food pigment in place of casein when applying a film on a piece of paper. Soy protein isolate is used as a protein-fortified raw material by enzyme-activated protein denaturation or by directly adding soy protein isolate to pastries, bread, and sausages. Soy protein isolate refers to powdered soybean products whose protein content is more than 90% in a dried base. This product should be free from fat, and its fibrous material content is characteristically very low.

Utilizing Soybean Protein Products

Recently soybean protein products such as defatted soy flour, soy protein concentrate, and soy protein isolate have been added to processed meat products like sausages and hamburger meat, to dairy products such as cheese and coffee cream, and to bakery products in order to fortify protein and reduce the cost of raw materials. However, effective use of soybean protein requires some considerations. Adding to relatively high-priced animal protein such as meat, eggs, or milk, or to bread, should not adversely affect the smell, flavor, texture, color or shape of such products. Among the soy protein products used for these foods, soy protein isolate is be the best choice as it affects the product's smell the least, and it is better if its storage period is short.

Processed Meat Products

For a long time in the West, relatively low-priced collagen or plant protein has been added to processed meat products in place of high-priced meat. Soy protein concentrate or texturized soy

protein is added to hamburger patty as an incremental raw material while defatted soy flour or soy protein concentrate is added to sausage products. Adding soybean protein this way can lower the price of meat, reduce shrinkage when cooked, and improve the food's emulsifying capacity and stability. It also increases the binding of meat particles and meat's moisture absorption capacity, improving the meat's texture in terms of enhanced mouthfeel when chewing the meat and the accompanying solid feel. Up to 25 to 30% of either texturized soy protein or soy protein concentrate can be added to hamburger patty, reducing the cost of raw materials by 20 to 30%.

How to Make Soybean Meat

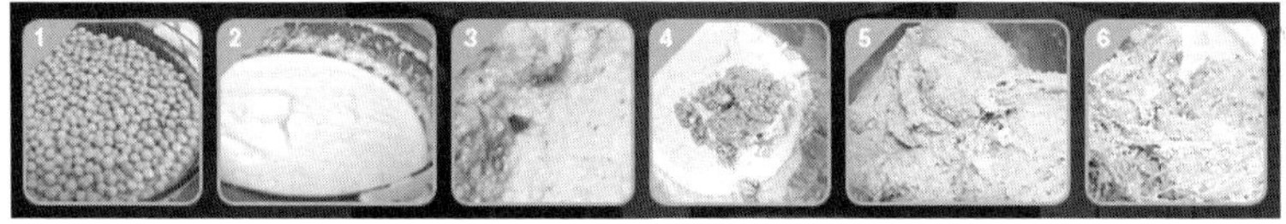

Making Soy Meat :

Raw Materials: One cup of soaked soybeans, 3 cups of water, a half cup of peanuts, a small amount of sesame, walnuts, and pine nuts, a spoonful of ginger juice, 3 spoonfuls of sesame or perilla oil, 2 spoonfuls of soy sauce, a spoonful of salt, a half cup of flour and a cup of gluten.

Method :

1. Grind the soybeans: Soak the soybeans for one night and then cook them slightly before grinding.
2. Grind the nuts: Grind the peanuts and other nuts using a grinder.
3. Mix the nuts and seasoning: Pour nuts into the bean soup, and add soy sauce, salt, and sesame oil for seasoning.
4. Make the dough: Make the dough by mixing the wheat flour with the gluten, adding the gluten little by little. Make sure the dough is not too thin or too thick.

Milk Products

Milk products to which soybean protein can be added include ice cream, cheese, coffee cream, whipping cream, and a mixed drink of soymilk and milk. For this mixed drink, linoleic acid, lysine, and phenylalanine, which are found in profusion in soymilk, are added to milk, resulting in a nutritive composition better than egg yolk while lowering the lactose content. In addition, beverages are available in which soy protein isolate is homogenized by dissolving in water, sugar, corn syrup, minerals, fat, vitamins, and an emulsifying agent.

Bread and Snacks

Adding soy protein products to bread or snacks can enhance not only the nutritional value of these products but also their moisture retention or absorbency. Adding 12% defatted soy flour can increase not only a bread's protein content by about 50% but also its protein efficiency. Soybean protein products added to bread and snacks help these products to enhance moisture retention, making it easier to knead flour into dough and resulting in softer crumb. Also it can expedite the browning of the bread crust and enhance a donut's oil absorbency, producing the overall effect of increasing nutritional value.

Texturizing Soybean Protein

Soybean protein products, such as soy flour, defatted soy flour, soy protein concentrate, and soy protein isolate, do not taste good eaten as is, thus necessitating the development of texturizing technology. Methods of texturizing soy protein include extrusion, spinning, and steam texturizing. Texturizing is accomplished by the breaking of the hydrogen bond in soybean protein, thus causing the protein to spread out, during the process of heating inside an extruder or during steam heating. The dispersed protein is rearranged lengthwise, and cooling such protein gives rise to the formation of a new hydrogen bond, which in turn precipitates the rearrangement of the protein molecules, enabling molecule crosslinks to form a stabilized texture. Soy protein products texturized this way are enjoyable and can easily be eaten, providing texture similar to meat in terms of flavor and chewing.

Utilizing Soybean oil

In the early years when soybeans were first introduced in Europe and the United States, soybean oil was produced by squeezing. However, demand for and consumption of soybean oil increased owing to the gradual development of society and population growth, necessitating the development of raw materials for mass production. The solvent extraction method was developed at just the right moment and opened the way for mass producing refined fat and oil out of the plant seeds at a moderate price. Soybean oil is used as a raw material for manufacturing shortening and margarine, whose production and supply are the greatest among fat and oil products. Also it is used as edible oil for roasting, frying, cooking, and in many other ways. The fat content of a soybean is about 20%, and the content of unsaturated fatty acid, i.e., oleic acid, linoleic acid, and linolenic acid, is about 85%, enabling it to exist in a liquid form at a wide range of temperatures and making it suitable for a variety of uses. It also contains plenty of tocopherol, a natural antioxidant, which contributes to oxidation stability and enhanced flavor. Its fat and oil can be converted from solid to liquid or vice versa depending on temperature, and for this reason it is used for pastries and bread making. Its potential use as shortening for frying is great. Production of soybean oil tends to grow in tandem with the growth of soybean production.

Uses for Soybean Oil

Fat or oil is heated at a constant temperature and then agitated after adding a catalyst (like nickel) with hydrogen gas to induce a hydrogen reaction. Hydrogenation quickly occurs, if various conditions such as temperature, pressure, type of catalyst, and amount of hydrogen are sufficiently met. Hydrogenation is accomplished if such a reaction continues until the double bonds of triglycerides are completely saturated. If that is the case, solid fat with a higher melting point can be obtained (hardened oil) and can be used for food processing. Typical hardened oil such as margarine and shortening is used for maintaining plasticity and enhanced oxidation stability, or they can be used as substitutes for butter.[1]

Manufacturing Soybean Oil Byproducts and Their Uses

Lecithin

To produce soybean oil, hot nucleic acids are extracted from soybeans at an earlier stage, and then filtered to remove minute particles, protein, and metal impurities. Nucleic acids are removed by distillation and the rough oil is treated washed with water to obtain insoluble fat that contains phosphorous. Emulsifying and moisturizing capacities are the desired qualities for the industrial applica-

1) Lee Gyeong-il, Soybean oil and its byproducts, Kong (Soybeans), Korea University Press, Seoul, 691-748(2005)

tion of lecithin. It is widely used for manufacturing livestock feed, bakery products, candy, chocolate, ice cream, macaroni, ramen, margarine, edible oil and fat, pigmented ink, and medicines.

Tocopherol

Tocopherol was first separated in 1922 by Herbert M. Evans. The word tocopherol comes from a combination of the words "tocos" (meaning offspring), "pherein" (meaning giving birth to), and "ol" (meaning alcohol). It is compared to childbirth because of the essential functions it performs for the body. Tocopherol works as an antioxidant and also functions as bioactive vitamin E, an essential nutrient. The extraction amount of tocopherol varies depending on the raw soybeans' ecological conditions, deodorizing process, which includes conditions such as deodorizing temperature and the deodorizing hour.

Defatted Soybean Cake

Mostly defatted cake is used for manufacturing livestock feed, and a small amount of soybean oil is used as an energy source for feed production. The history of using soybean meal as a raw material for livestock feed coincides with the history of soybean oil extraction. In the early years, bean cake, a byproduct of the squeezing or extracting method, was used as animal feed without any further processing. However, a new technique was introduced in the 1930s, in which anti nutritional factors were inactivated by toasting soybean meal, leading to the creation of soybean meal with a higher nutritional value. Soybean oil meal, soybean hulls, and soybean lecithin are produced as byproducts during the processing of soybean oil squeezing, soybean oil refining, and soybean protein refining.

Uses of soybeans other than for food in Korea

- Defatted Soybean Cake: Bean cake is a byproduct from the process of extracting soybean oil from the soybean and is used as an important raw material for manufacturing livestock feed. In the past, an animal protein source was mainly used as the livestock feed's protein source. However, it grew impossible to meet demand with the animal protein source alone. For this reason, soybean cake began to be utilized, as its supply was relatively stable, it was of good quality, and was available at a moderate price. Soybean cake was also used for catching fish in streams.
- Fertilizer: Soybeans are left growing in a farrow field, and when the field is plowed, the bean plants are turned under to nourish the soil. Then, when weeding the field, scatter 4 *mal* of soybeans (1 *mal* is equivalent to about 18 liters) in 660m^2 though 1 to 2 *mal* will suffice if 4 is not practical. During the summer, these scattered soybeans decompose and serve as fertilizer.
- Detergent: Water strained off coagulating tofu can be used as detergent for washing hands or taking a bath. Sometimes frostbitten hands or feet can be soaked in it.
- Red stamp ink: It was said that red dye mixed into soymilk could become a long-lasting red stamp ink.
- Raincoats: At first sesame oil is applied and the garment is dried in the shade, then eggs, talcum powder, pine resin powder, perilla oil, and soybean are finely ground and mixed, and then applied 4 or 5 times to the inside and outside of the raincoat.
- Waxing the floor paper: White soybeans are soaked in water at a ratio of 1 to 6 and then finely ground and wrapped, using hemp cloth to squeeze. The bean juice made this way is mixed with perilla oil at a ratio of 4 to 1. The ratio of 2 to 1 should be used when applying it to a wooden floor.
- Small kerosene lamp: Wicks were made by twisting *hanji*, Korean paper hand-made from mulberry trees, with cotton wool and flax, and soybean oil and castor oil were used as fuel. Lamps with twin wicks emitted very bright light but consumed much fuel.

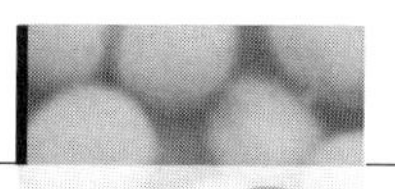

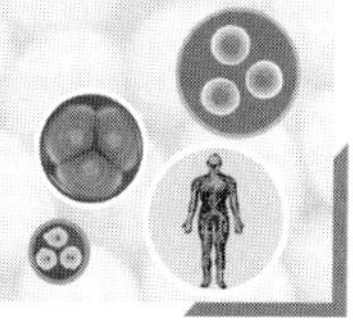

Nutrition and Functionality of Soybeans

Chapter

05

1. Structure and Composition of Soybeans

In the 1970s in the United States an alarm was sounded about modern human food consumption choices and eating patterns. At that time a special committee was set up in the US Senate to investigate America's eating habits. The study lasted two years, and the final report came to about 5000 pages. At first the committee was interested in issues of poverty and hunger, but soon it gravitated towards food and nutrition, or how eating habits affect health. In January 1977, committee president George McGovern said, "The clear fact is that our eating habits over the past half century have changed for the worse, resulting in a profoundly negative impact on our health. The excessive intake of fat, sugar, and salt correspond directly with various fatal diseases, such as heart disease, cancer, and stroke. Six out of ten of fatal diseases in America are related to diet." The release of this committee's report triggered a well-being craze in America and around the world as people took interest in their diets.

Structure and Composition of Soybeans

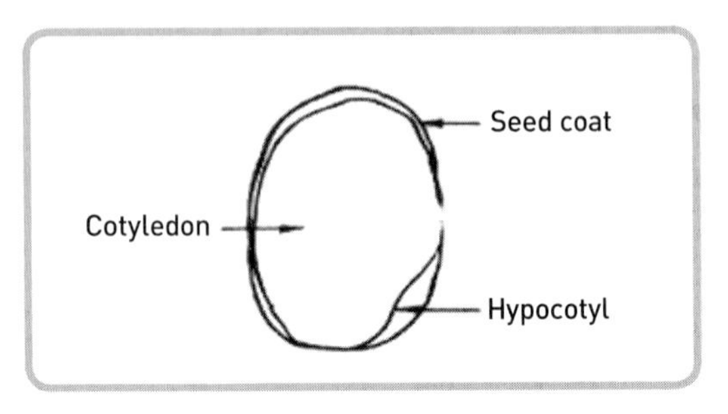

Composition of a Soybean

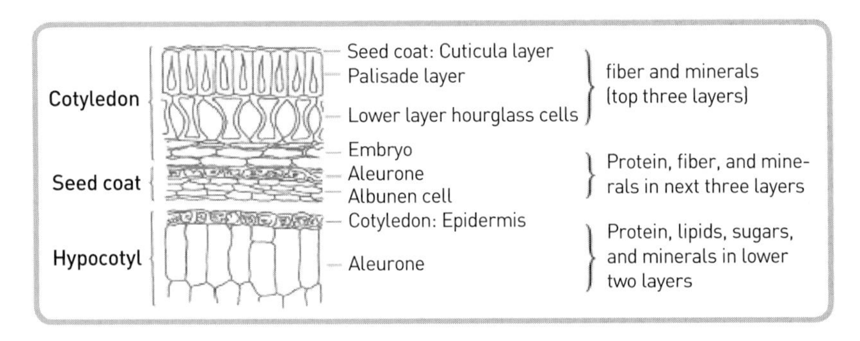

Composition of the Inside of a Soybean

The Structure of Soybeans

When it is time for a soybean plant to bear fruit, seeds are made inside the developing seed pod. Soybean seed consists of a seed coat, cotyledon (seed leaves), and a hypocotyl. Looking at each part separately, the seed coat weighs about 8.3% of the entire seed, the hypocotyl comes to 2.1%, and the remaining 90% belongs to the cotyledon. This composition ratio differs according to the size and breed of the soybeans. The seed coat is mostly (86%) made up of fiber and soluble nitrogen, while the edible cotyledon contains over 40% protein (in dry matter) and about 20% lipids. This is the part that will develop into the first leaves. The point at which the first leaves divide is called the growth point, and from there the stem and leaves will emerge as the plant sprouts. The hypocotyl becomes the rhizome and then the root.

2. Soybean Nutrition and Bioactive Substances

Among Westerners with a high intake of animal foods, the main cause of death comes from the formation of chronic degenerative diseases. Conversely, East Asians have a lower morbidity rate vis-à-vis these diseases. It is possible that this difference is related to the fact that there are a large number of soy foods in the East Asian diet. Compared with other vegetable products, soybean foods contain high levels of lipids and, when used as a substitute for meat or dairy products, the saturated fat and cholesterol content is much lower, thus making soy an excellent health food.

Soybeans contain a balance of three nutrients: protein, fat, and carbohydrates, at a rate of about 40%, 20%, and 18%, respectively. Beyond that, soybeans contain vitamins, minerals, and fiber, and recent research has shown that they also contain isoflavones, lecithin, and other phytochemicals. Aside from water, the main ingredient in our body is protein. The greatest value of the soybean as food is its role as a protein source. In the West from early on, soybeans were sometimes referred to as "field meat." Soybeans have two and a half times the amount of protein as meat. As proteins from soy decompose they produce peptides, and as these decompose, they produce amino acids. Not only the proteins in soy, but also the effects of peptides on the body have been objects of recent research. Soy protein is known to reduce blood cholesterol levels and the risk of blood vessel disorders such as arteriosclerosis and heart disease.

Just How Good are Soybeans?

In the past the protein score was the evaluation method used to determine the nutritive qualities of protein. Animal protein was considered superior, and vegetable protein was evaluated as inferior. Recently, however, a new method called Protein Digestibility-Corrected Amino Acid Score (PDCAAS) has been used to find the value of proteins by testing the volume and rate of necessary amino acids found in protein, as well as the rate of digestibility and other affective elements. Calculations based on this method show that soybean belongs to the pool of foodstuffs with excellent-quality protein (see diagram below).

Healthful unsaturated fatty acids make up over half the lipids in soybeans: Of the total amount of lipids, linoleic acid (omega-6 fatty acid) comprises 57%, oleic acid (omega-9 fatty acid) 23%, and linolenic acid (omega-3 fatty acid) 8%. Omega-6 and omega-3 fatty acids are necessary elements for our bodies that we cannot produce ourselves but must obtain from foods, and it has been shown that omega-6 fatty acids in particular are effective in suppressing the growth of cancer cells. Not only do these fatty acids directly suppress the growth of cancer cells, but they also induce the cancer cells to self-destruct.

Essential amino acid composition and chemical score of Soybeans, Milk, Beef, and Polished Rice (mg/gN)[1)]

Essential Amino Acids	Reference	Soybeans	Polished Rice	Beef	Milk
Isoleucine	270	336	322	327	407
Leucine	306	482	535	512	626
Lysine	270	395	236	546	496
Phenylalanine	180	309	307	247	309
Methionine	144	84	142	155	156
Threonine	180	246	241	276	294
Tryptophan	90	86	65	73	90
Proline	270	328	415	347	438
Chemical score	100	73	72	83	78

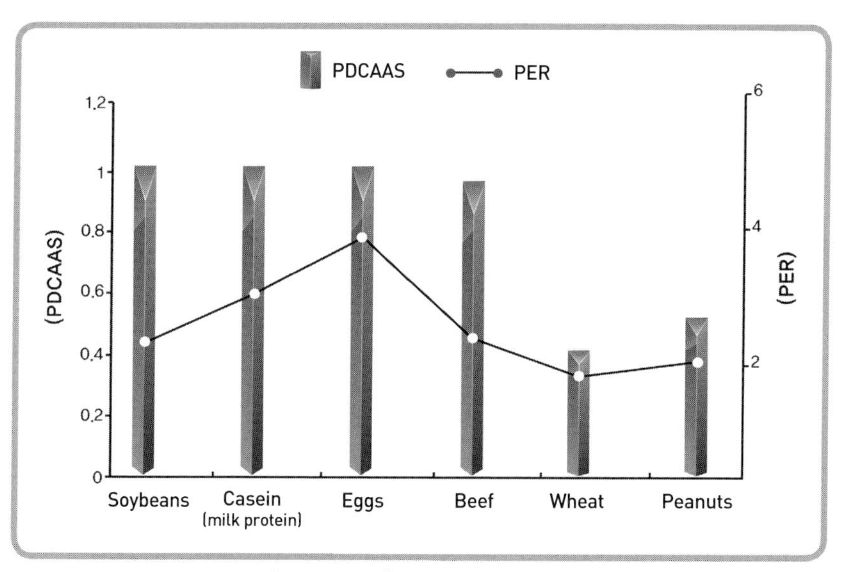

Comparison of PDCAAS (method considering the rate of digestion of amino acids) with Protein Efficiency Ratio (PER)

1) Kim Woo-jeong, The processing characteristics of soybeans, *Kong* (Soybeans) Korea University Press, Seoul, 217-312(2005)

The representative sugar in soybeans, oligosaccharide, is composed of stachyose (3-4%), sucrose (about 4%), and raffinose (about 1%). Oligosaccharides are a kind of carbohydrate that, as a composite of about 2-10 simple sugars, play a part in the proliferation of the healthy bacteria bifidus in the human body. Putrefactive bacteria are excreted after absorbing the oligosaccharides in soybeans, and lactic acid bacteria as well as short-chain fatty acids such as acetic acid and pyruvic acid, which are metabolites of oligosaccharides, have a cleansing effect on the intestines.[1)]

Half a cup of boiled soybeans contains 44% of the recommended daily allowance (RDA) of iron, as well as a significant amount of calcium, magnesium, and zinc. Soybeans are also a good source of B vitamins, containing vitamins B_1, B_3, and B_6. Unlike animals, humans do not have an enzyme that digests fiber. Among fibers in soybean, insoluble fiber promotes interlocking action in the intestines and absorbs heavy metals and other toxic substances to be excreted together with them. In this way they aid in preventing constipation and colorectal cancer. On the other hand, soluble fibers prevent rapid increases in blood sugar, thereby helping prevent diabetes.

Soybean bioactive Substances

Isoflavones

Isoflavones are substances that protect the plant when it is threatened by the outside environment. Soybeans have more isoflavones than any other food, which is a plus because soybeans can so easily comprise part of one's diet. Isoflavones are also called phyto-

---

1) Seung Jeong-ja, Nutritional value and functionality of soybean foods, *Kong* (Soybeans), Korea University Press, Seoul, 579-660(2005)

estrogen, where "phyto" means plant and "estrogen" refers to the effects of isoflavones on the body, which are similar to those of the female hormone estrogen. Depending on the situation, isoflavones may have a regulatory role on the effects of estrogen. If the female hormone levels are low, the isoflavones in soybeans can substitute for the role of the hormones. If the hormone levels are too high, the isoflavones regulate the volume. Isoflavones also raise bone density levels and make the bones stronger. The RDA of isoflavones has not yet been determined, but 50-100mg per day is suggested. In one cake of soybean curd there is 150mg, in 200ml of soymilk there is 30mg, and in 15g of *doenjang* (fermented bean paste) there is 5.5mg.

Lecithin

The combination of phosphorous and lipids is called phospholipids. In the body lecithin, a phospholipid, creates cell membranes, emulsifies cholesterol, and helps cholesterol produce hormones. Lecithin is also a source of the neurotransmitter acetylcholine, which plays an important role in enhancing memory.

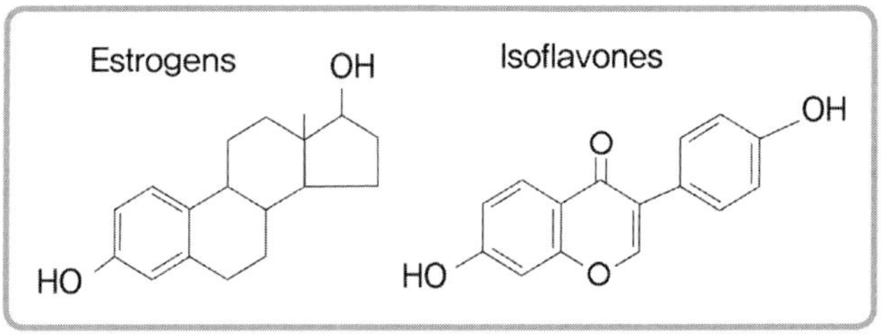

Estrogens OH HO Isoflavones OH O HO O

Molecular structure of Isoflavone and Estradiol[1]

1) Kenneth Setchell, Phytoestrogens: The Biochemistry, Physiology, and Implication for Human Health of Soy Isoflavones, *American Journal of Clinical Nutrition*, 68, 1333S-1346S(1998)

Saponin

When soybeans are soaked and boiled in water, white bubbles appear; these bubbles are made up of saponin. Saponin regulates cholesterol by blocking the absorption of cholesterol in gastro-intestinal track and is also known as a functional substance that delays aging. The characteristics of saponin is to effectively dissolve well into water and oil, thus suppressing the formation of lipoperoxide, which is made in the process of lipid oxidation, and lowering the rates of cancer and arteriosclerosis.

Polyphenols

The polyphenol which shows antioxidant and anticarcinogenic properties, is a phytochemical found in soybeans and many other types of plants. Polyphenols have the function of cleaning any chemical garbage that enters the body. They disrupt the chemical ingredients that may lead to the formation of tumors and suppress the growth of cancer cells in the body.

Phytic acid

It has been reported that phytic acid prevents the absorption of iron, but recent research shows that it rather binds with iron in the body to maintain iron content, thus suppressing free oxygen radicals. Phytic acid is particularly effective in preventing colorectal cancer.

Pinitol

As a carbohydrate derivative, pinitol is known to have insulin-like behavior in the body. Due to the variable absorption rate of pinitol, its absorption efficacy has been investigated in diabetic patients. Pinitol has been researched as capable of preventing, without side effects, the rapid increase in blood sugar after a meal, as well as having a possible effect on the prevention and curing of diabetes.

3. Functionality of Soybean Products

Soymilk

Nearly all the ingredients contained in soybeans are also in soymilk. From long ago, the people of Korea used the words "soybean soup" to indicate a summer drink or a soup with soymilk and noodles, but now soymilk itself is used as an alternative to milk. Soymilk is an important source of biomembrane-building polyunsaturated fats and fat-soluble vitamins. It contains the antioxidant nutrient vitamin E at a level of 93.7mg per 100mg of soymilk, which has the effect of inhibiting cell aging. Not only does soymilk contain the protein ingredients of its parent plant, but it also contains bioactive substances such as dietary fiber, oligosaccharides, saponin, and isoflavones. And because soymilk contains no lactose, it has great value as a source of full nutrients without side effects for those who are lactose intolerant[1] or allergic to milk.

Soybean curd

Among soybean ingredients, lipids and protein transfer best into bean curd and soymilk. Typically bean curd contains 80-85% water, about 8.5% protein, about 5.5% lipids, and about 1.5% sugar. These nutrients, which are absolute necessities for the body's metabolism and growth, include essential amino acids and fatty

1) Lactose intolerance occurs when the lactose in milk cannot be decomposed by the body, leading to symptoms of indigestion. About 70% of Koreans are lactose intolerant.

acids, as well as calcium, iron, and other inorganics, of which bean curd, as a protein plant food, contains much.

Soybean sprouts

In the process of soybeans germinating and sprouting, their internal metabolism causes their nutritive ingredients to change significantly. During the growth process, lipids within the seed decrease significantly, while dietary fiber increases and vitamins increase in great volumes. For example, soybeans contain no vitamin C, but in the process of sprouting, about 16mg of vitamin C in 100g of soybean sprout is created. Vitamins K and niacin are also created in the sprouting process. By oxidizing, lipids decrease the amount of toxic lipoperoxides that are formed; the amount of vitamin B_2, which functions to prevent the hardening of arteries, doubles; and dietary fiber, which helps prevent obesity and constipation, also increases in content volume. Compared with other pulses, soybean sprouts contain more iron and protein.

The amino acids in soybean sprouts include essential amino acids (lysine, tryptophan, methionine, etc.) and free amino acids, so that when a person is worn out, eating these sprouts can have a positive effect on restoring function. In the rootlets of soybean sprouts, asparagine (an amino acid) helps relieve hangovers due to its aid in the function of enzymes breaking down alcohol. The potassium in soybean sprouts has the opposite effect of sodium in that it helps to lower blood pressure.

Soybean oil

High in polyunsaturated fats, soybean oil provides an important source of antioxidant nutrients such as vitamin E (tocopherol). More than any other plant oil, soybean oil has essential fatty acids and vitamin E, thus providing a balance mix of nutrition and functionality. Per 100g the oil contains 55% or more of essential fatty acids, including polyunsaturated linoleic (omega-6) and linolenic acid (omega-3). Omega-6 fatty acids comprise 52.6%, omega-3s, 7.94%. The ratio of omega-6 and omega-3 fatty acid composition in soybean oil is 8.21 : 1, which achieves the nutritional balance suggested in countries around the world for recommended intake, which stands at a ratio of 4-10:1. Research has shown that the polyunsaturated fats in soybean oil lower total and LDL cholesterol, thus becoming a support to one's overall health.

■ *Doenjang* (Fermented soybean Paste)

The longer *doenjang* ferments, the more free fatty acids, isoflavones, and omega-6 fatty acids increase, thus expanding its bioactive effect. In the fermenting process, the browning compounds that are formed (including amino acids, peptides, and proteins in the amino group, as well as reducing-sugars such as glucose and the colored compounds known as melanoidins that begin to appear during the Maillard reaction) and several types of phospholipids, tocopherol, amino acids, peptides, and so on, contribute to its anti-oxidative function. A strain of Bacillus in *doenjang* secretes subtilisin (a protein-digesting enzyme), and traditional *doenjang* contains 10 times more subtilisin than commercial *doenjang* and 3-4 times more than *cheonggukjang* (a fast-fermented soybean paste) and *natto* (Japanese soybeans fermented with *Bacillus subtilis* var. *natto*). Bacillus bacteria generate organic acids that have a regulating effect on the intestines. Bacillus spores can penetrate the intestines and germinate there, becoming vital probiotics in the gut. Not only do the omega-6 fatty acids contained in *doenjang* have a suppressing effect on the growth of cancer cells, but they also have an anti-tumor effect by inducing cancer cells to self-destruct. Due to *Bacillus* and other probiotics that exist in fermented soybean products, these foods are excellent for cancer prevention and antioxidative function.[1)]

■ *Cheonggukjang* (Fast-fermented soybean Paste)

Cheonggukjang borrows the strength of bacteria to break down proteins. Among fermented soybean foods, this one is complete in the shortest amount of time. *Cheonggukjang* is fermented using *Bacillus subtilis*, and per 100g, 100 billion digestion-improving enzymatic bacteria form. In addition, during the process of fermentation beneficial bioactive substances that were not part of the original, healthful ingredients of soybeans are formed, thus increasing the excellence of this native food. The main role of bacillus in *cheonggukjang* is to weaken the putrefying bacteria in the intestines and act as an antibiotic to pathogenic bacteria. A substance called polyglutamic acid is produced from the sticky mucilage of *Bacillus subtilis*, this chemically tasty product is a glycoprotein formed by the combination of about 5000 amino acid called glutamic acid, connected peptides and fructose. *Cheonggukjang* helps with prevention and cure of osteoporosis due to its ability to help the body expedite the absorption of calcium, and it promotes intestinal regulation while reducing intestinal gas. It also

---

1) Park Kun-young, Health functional properties of fermented soybean products, *Kong* (Soybeans), Korea University Press, Seoul, 407-454(2005)

contains the bioactive substances such as isoflavones, phytic acid, saponin, trypsin inhibitors, tocopherol, unsaturated fats, dietary fiber, oligosaccharides, antioxidants, subtilisin, vitamin K, and more. In order to obtain the most enzymes, eating it raw rather than boiled will be most efficacious. However, as there are many other healthy ingredients in *cheonggukjang*, eating it in stew is also beneficial.[1]

Gochujang (Fermented Red Pepper Paste)

Gochujang contains 10% soybeans (in the form of *meju*, fermented and dried soybean). Although *gochujang* is the latest traditional sauce to be introduced, it is the first internationalized Korean sauce. The spicy taste of *gochujang* comes from the ingredient capsaicin, which along with the colorant capsanthin (a carotenoid) are additional healthful ingredients supplementing the soybeans. Nutrients derived from *meju* help prevent cancer, the capsaicin in red pepper powder aids weight loss and improves endurance, and capsanthin has an antioxidant effect. Capsaicin also stimulates the central nervous system via the blood and promotes the secretion of adrenalin from the adrenal cortex via the sympathetic nerves, and because it causes fat to burn, its ingestion promotes the metabolism and energy levels nearly to the extent of having exercised. In this way it can also be a preventative against obesity.

1) Kim Han-bok, *Cheonggukjang daieoteu & geongangbeop* (*Cheonggukjang* diet and healthy life), Human&Books, Seoul(2003)

Increasing Effectiveness through Fermentation

In making soy sauce or *doenjang* using soybeans and rice, or perhaps flour, the fermentation process creates a delicious and pleasing taste from the amino acids that occur when the protein within soybeans is broken down by the enzyme produced by the mold that grows on *meju*. The starches in rice or flour that are fermented with soybeans break down into sweet glucose or maltose. The combination of the amino acids with these sugars creates a unique taste, to which *gochujang* adds hot spice for a deep, savory seasoning. In this way, the microbes in fermented foods break down the original ingredients of a food, but when the materials of that decomposition are combined with other ingredients the character of the food becomes another thing entirely.

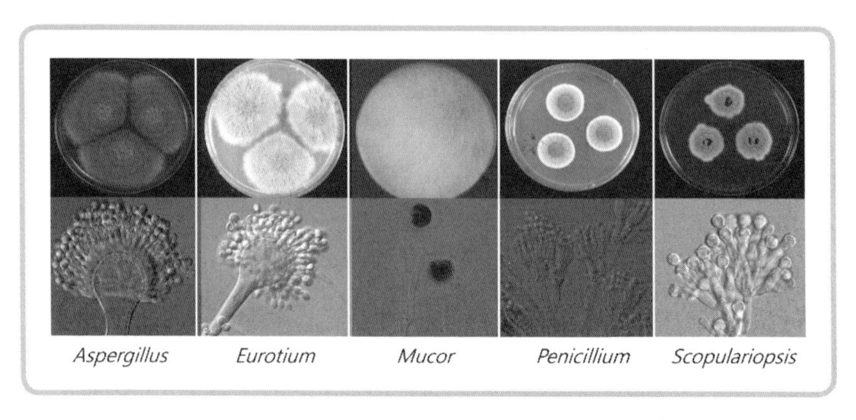

Microbes Involved in Soybean Fermentation[1)]

1) Hong Seung-beom, Personal Communication, National Academy of Agricultural Science, Wanju, South Korea(2011)

Increased Digestion-absorption Rate

The digestion-absorption rate for soybeans is 65%, for bean curd and *doenjang*, 85%, and for soymilk and *cheonggukjang*, 95%. The nutritive qualities of soy proteins change dramatically in the process of heating or fermenting. The quality of the protein increases greatly due to the fact that the trypsin inhibitor in soybeans, which normally would block the function of the enzyme that breaks down the protein, is rendered inert or dissolved in the heating or fermenting process. The isoflavones contained in soybeans are known as bioactive compounds that perform the function of preventing osteoporosis and cancer. In food, isoflavones bonding chemically with sugars are glycosides, or if isoflavones do not combine chemically with sugars they become aglycones. Aglycones are known to be more healthful for use in the body. In bean curd and soybean sprouts, similar to full-grown soybeans, most isoflavones exist in the form of glycosides, but in fermented foods such as *doenjang*, a high volume of isoflavones break down to become much more bioavailable aglycones in the body.

Increase in vitamins

In raw soybeans there are not many B vitamins, but in *doenjang* or *cheonggukjang*, which are bacteria-fermented, B vitamins abound. Vitamin C is not found in soybeans, but it is plentiful after adding water to seeds and sprouting them. If the number of homocysteines (an amino acid not found in protein) in the blood increases, it is highly likely they will clog arteries, but vitamins B_6 and B_{12} lower the number of homocysteines in the body. Vitamins in the B group also play an important role in breaking down sugars and converting them to energy. Accordingly, it is important to take plenty of B vitamins when hoping to invigorate the brain. Among the key nutritive elements for improving memory are vitamin B_3 (niacin), vitamin B_5 (pantothenic acid), vitamin B_6 (pyridoxine), and vitamin B_{12} (cobalamin). Vitamin B_3 in particular demonstrates great power in improving memory function.

The Effects of Probiotics

"Probiotics" is a generic term for all the bacteria that are good for our bodies, such as bifidobacteria and lactobacillus. One representative strain of beneficial bacteria that resides in the intestines is lactobacillus, and as it fights harmful bacteria in the gut it secretes antibiotics. Although lactobacillus was first isolated in milk, whence its name, this strain is plentiful in *doenjang* and *cheonggukjang*. The lactobacillus in *doenjang* can withstand salinity, surviving in a salt water solution of about 20%, and it is also heat-resistant up to 100℃. Korean food development researcher Choe Sin-yang has confirmed that *doenjang* extract causes a proliferation of B lymphocyte, which reinforces the immune system. That this immunity-regulating agent has not been found in raw soybeans, but is found in *doenjang*, testifies to the variety of microorganisms that exist in the natural world due to fermentation.

Antioxidants and Brown Pigment Formation

The active oxygen circulating in our bodies damages our cell membranes, DNA, and other cell structures, and depending on the scope of that damage, our cells can lose their function or become altered in detrimental ways. The secret to the antioxidant effect of *doenjang* is hidden within its brown pigment formation. As *doenjang* ferments, the water soluble and fat soluble brown pigmentation that is created contains vital antioxidants. This brown pigment prevents unnecessary physiological oxidation and thus is effective in aging prevention. The results of studying the browning agents created in *meju* and *doenjang* as they ferment, and the effects these browning agents and phenol compounds have on oxidation stability in lipids, show that as *meju* ferments, at first the amount of water soluble browning agents is low, but as fermentation progresses the amount of these agents increases dramatically, and as *doenjang* ferments there is a continuous increase in these agents until they reach a high-level content. In *meju* the content of phenol compounds increases at the beginning of fermentation and reduces somewhat towards the later stages of fermentation, but the water soluble browning agents and phenol compounds, with respect to the omega-6 fatty acids, demonstrate a strong antioxidant effect. When measuring the brown pigment in extract of soy sauce-brewed *doenjang*, the amount increased with the length of fermentation time, and water soluble agents of brown pigment had a greater antioxidant effect than those that were fat soluble. The results of these tests show that even while raw soybeans contain many body-healthy ingredients, *doenjang* contains a higher amount of functional materials than raw soybeans.

Thrombolytic Enzymes

The most representative functional materials in soybeans are the thrombolytic enzymes that suppress strokes. These enzymes dissolve blood clots that form when blood vessels stack up and impede blood flow through the arteries, which can then lead to heart disease or stroke. There are no thrombolytic enzymes in raw soybeans, but they have been discovered in *doenjang* and *cheonggukjang*, which have undergone the process of fermentation. In addition, it has been confirmed that during the lipid oxidation process in *doenjang* the formation of peroxides is suppressed, and there are also antithrombocytic peptides, which actively decrease the agglutination of platelets.

Immunity Boosters

Among the results of ingredient-testing in *doenjang*, research has confirmed that in *doenjang* extract there is a proliferation of immunity-boosting B-class lymphocytes. Also, the formation of macrophages (immunity-boosting cells) and cytokines (immunity regulatory cells) in B-class lymphocytes is increased in *doenjang*, thus boosting immunity. Long ago *doenjang* was spread on people's skin where they had received a bee sting or snake bite; it was known to have a regulatory effect on genes connected with infection or inflammation. These kinds of immunity-regulating agents have not been found in raw soybeans, but can be found in *doenjang*. These ingredients are the functional substances grown in the process of making *doenjang* that transform to boost the immune system according to the various actions of the microbial products of fermentation.

Soybeans and Nutritionally Compatible Foods

Any one given food cannot provide sufficient nutrients to fill all of a person's needs in a day. Although soybeans offer a very full nutrient profile, it is not complete. Listed below are foods that complement the nutrients in soybeans, so-called compatible foods, to offer a nutritionally complete meal.

Kongbap: Soybeans with Rice

Although rice is high in starch, it is low in protein and lipids. Among essential amino acids, it is deficient in lysine and tryptophan, while methionine abounds. Compared to this, soybeans have little starch but are rich in protein and lipids. Soybeans are low in methionine. On Korean dinner tables soybeans and rice have shared a long history of supplementing each other's weak points as complementary foods. Habits of eating soybeans chopped into rice cake, soybean and noodle soup, soybean-powdered rice cake and other dishes help supplement rice with needed protein.

Kongguksu: Soybeans with Wheat

The best healthy food for hot summer days when it is easy to sweat a lot and not have much appetite is soybean soup. In wheat flour the essential amino acids that make up proteins such as lysine, methionine, threonine, and tryptophan are in short supply, and soybeans contain 3-5 times the amount of these amino acids than flour, thus achieving a supplementary effect, and soybeans also have a good supply of vitamins B_1 and B_2, which flour lacks.

Licorice Bean Soup: Licorice with Black Soybeans

Licorice has a medicinal quality that helps the urinary and lung systems and it is effective against coughs. In addition, it can lower a fever, counteract poisons, and help new skin to appear. Black soybeans help the body quickly rid of toxic substances and break down other noxious agents to ease the overall effect of toxins on the body. Licorice bean soup is used for its strength in detoxifying the heavy metals that can build up in the body.

Doenjang Stew: _Doenjang_ with Vegetables

A food that supplements the sodium present in *doenjang* would be one that is full of potassium. To make *doenjang* stew, one boils the *doenjang* with anchovies and kelp and then adds Chinese chives, potatoes, zucchini, and shiitake mushrooms. Chives in particular are high in potassium, have a distinctive smell, and are known to aid digestion. Boiling chives in *doenjang* soup is effective against diarrhea. Chives also help strengthen the intestines and are good for those whose bodies run cold.

Miyeokguk (Seaweed Soup): Seaweed with Soy Sauce

In Korea people eat *miyeokguk* on their birthday and after giving birth. There are some African tribes that eat seaweed soup, but in no other country is it eaten every year on birthdays as well as on the day when a baby is born. One story goes that people who lived by the sea noticed that whales would pick and eat seaweed when it came time to calve, and the people followed suit. The saponins in soybeans are a good ingredient for fighting cancer, but when they enter the body they also discharge iodine outside the body. Thus it is always good for one who eats soybeans to also ingest foods with plenty of iodine. Seaweed contains plentiful iodine, so consuming seaweed soup with soy sauce provides a well- balanced meal. Beyond *miyeokguk* there is a variety of other, similarly healthy dishes such as seaweed with vinegar, soybeans boiled in soy sauce with kelp, soybean curd and seaweed seasoned with vinegar, and *doenjang* stew made with kelp broth.

Jo Han-yeong 「*Tongsok hanuihak wollon*」

In 1934 in *Tongsok hanuihak wollon* (Principles of Korean folk medicine), Jo Han-yeong explains the tradition of women eating seaweed soup after giving birth as follows: "In Joseon we give women in childbirth as much seaweed soup as possible." Explained in light of herbal medicine, seaweed has sweet, strong, and slightly sour tastes, and the color can be black, green, or yellow; and whether because of taste or color, it served as medicine to regulate the three foot meridians, that is, the foot meridians that radiate upwards as the bladder meridian, the liver and gall bladder meridian, and the spleen and stomach meridian. In any case, in modern times seaweed has been medically shown to contain a large amount of iodine, and brown algae (or *miyeok*, that which is used in traditional seaweed soup) contain a particularly high amount. Iodine has the following effects on the body, making seaweed singularly suitable to women who have just given birth: 1) It stimulates the metabolism, 2) disinfects blood cells by destroying bad bacteria therein, and 3) neutralizes and eliminates toxins within the body. The prescription for eating seaweed soup post-partum is the product of the people's experience in Korea, and will not be found in any current medical or traditional herbal remedies textbook. In his book *Seongho saseol* (*Seongho Miscellany*), Seongho Yi Ik said, "Heat governs menarche, and the fact that seaweed is good medicine for post-partum women is an important cultural prescription."

Creating a Healthy Dinner Table for All Stages of Life Using Soy Products

Great efforts are being made to solve the nation-wide problem of Western disorders currently increasing in Korea due to over-indulgence in animal food products. However, the solution lies very near, in the materials used in Korean traditional foodways. Soybeans themselves and soybeans in the form of fermented sauces, as well as soybean curd and other forms used in cooking and processing have been consumed for a very long time in Korea. Unlike Westerners, Koreans are accustomed to eating soy products in their daily diet, so it only remains to either maintain or increase the amount of intake.[1)]

A great variety of soy foods have been developed over time in Korea. Looking at each of the life stages, it has been judged that although total calories for those eating a significant soy diet are similar to those without, lipid and cholesterol intake is low, and calcium and isoflavone intake is high. According to a study of centenarians in Korea, their diet consisted of a traditional Korean diet, and the common characteristic among healthy older people was that they enjoyed eating soybeans, fermented foods, and natural foods.

Diet for Pregnant and Nursing Women

Substituting soy products into a nursing mother's meals will prove to be nutritionally compatible with her normal diet due to the increase in calories, iron, calcium, and protein.

- Cooked rice with soybean (*kongbab*), seaweed soup (*miyeokguk*), soybean pancake (*kongjeon*), panfried driedsardin and perilla leaf (*Myeolchiggatipbkeum*), seasoned soybean sprout and chive (*kongnamulbuchunamul*), watery kimch with radish and cabbage (*mabakkimchi*).

Diet for Age Groups 20-49 and 50-64

Even replacing animal protein with the plant protein of soy products, protein and calorie intake will not diminish when compared with a typical diet, but there will be an increase in the preventative effect against adult diseases.

---

1) Seung Jeong-ja, The nutritional value and functionality of soybean foods, *Kong* (Soybeans), Korea University Press, Seoul, 579-660(2005)

- Cooked rice with soybean (*kongbap*), soybean curd dreg stew (*kongbiji jjigae*), tri-colored chilled vegetables (*samsaek naeng-chae*), seasoned acorn jelly salad (*dotorimuk muchim*), seasoned aster bean paste salad (*chwinamul doenjang muchim*), and young radish *kimchi* (*yeolmu kimchi*).

Growing Children and Adolescents

Substituting soy products in a child's diet supplies enough necessary protein for the proper growth of body tissues. This is an important period for the development of the brain as well, and a soy diet supplies the needed calcium, much more so than a standard diet.

- Cooked brown rice with soybeans (*hyeonmi kongbap*), tofu radish soup (*dubu muguk*), seasoned soybean sprouts (*kongnamul*), savory soybean pancake (*kongjeon*), and seasoned cucumber and onion (*oi yangpa muchim*).

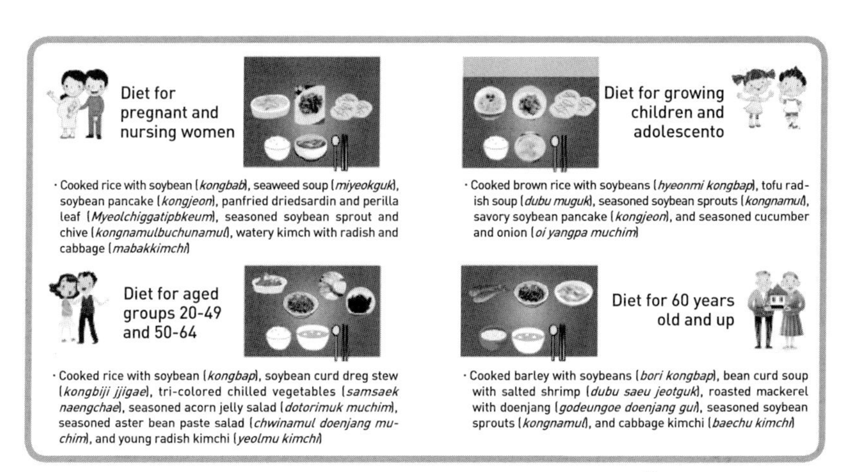

Healthy dinner table for the stager of life[1]

1) Seung Jeong-ja, The nutritional value and functionality of soybean foods, *Kong* (Soybeans), Korea University Press, Seoul, 579-660(2005)

Diet for 60 Years Old and Up

At this stage of life there are many instances in which protein and vitamins and minerals must be carefully managed in order to prevent nutritional deficiencies. A traditional diet using soy foods is advisable at this time.

- Cooked barley with soybeans (*bori kongbap*), bean curd soup with salted shrimp (*dubu saeu jeotguk*), roasted mackerel with *doenjang* (*godeungoe doenjang gui*), seasoned soybean sprouts (*kongnamul*), and cabbage *kimchi* (*baechu kimchi*).

4. Efficacy of Soy Foods on Various Diseases

Because soybeans contain several functional materials, they are used for the prevention and cure of diverse diseases. In the past people even used soybean and fermented soybean products as traditional folk medicines. *Bencao gangmu* (Compendium of Materia Medica), written during China's Ming dynasty states, "Boiled soybean juice draws out the body's toxins. It regulates kidney disease, makes urination productive, and lowers the energy. It prevents fever and chills, helps the blood circulate vigorously, and draws out all toxins." In *Dongui bogam* (*Principles and Practice of Eastern Medicine*), Heo Jun writes, "If bitten by an insect, mix burnt powder of black soybeans in water to make a paste and spread it on the bite to detoxify it. Also, a tea of this will cure the five viscera and gastrointestinal lumps, all eruptions, severe thirst, constipation, and the inability to urinate." The recently compiled *Hyangyak daesajeon* (Encyclopedia of traditional Korean medicine) states that this tea "helps the blood flow smoothly, has a diuretic effect, helps prevent stroke, detoxifies the system, cures jaundice edema, abdominal dropsy, etc., and weakens pathogens." It is known even today among the people that black soybean vinegar is good for gout.

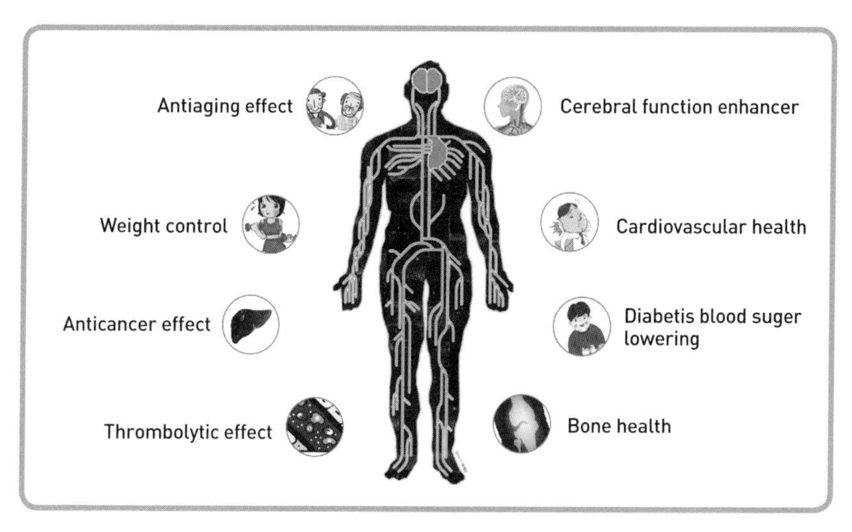

Physiological function of soybean[1)]

Soybeans Good for the Prevention and Cure of Diseases

Soybeans are Good for Cancer

Cancer cells occur when normal cells for whatever reason (oxidation, lack of exercise, stress, etc.) mutate, and the resulting abnormal cells continue to divide until they accumulate and change the structure of an organ and metastasize into different tissues. When cancer metastasizes it inflicts severe harm on the body. One method of prevention is to block the process of oxidation by ingesting antioxidant

1) www.soyworld.org

foods, of which soybeans are an outstanding choice. Anticarcinogenic ingredients in soybeans and fermented soybean products include isoflavones, polyphenol compounds, saponins, omega-6 fatty acids, tocopherols, phytic acid, peptides, and the browning substances created during the process of fermentation. The anti-tumor function of isoflavones is receiving particular attention in the scientific world. Isoflavones have a structure chemically similar to the female hormone estrogen and can combine with estrogen acceptors, thereby working on the principle of blocking the natural estrogen acceptors from beginning the division of mammary cells. Also, one isoflavone only found in soybeans, genistein, is effective in inhibiting the progress of prostate cancer. Among fermented soy products, the antioxidant activity of *cheonggukjang* (fast-fermented bean paste) is particularly high, containing eight times the antioxidant activity of soybeans. Numerous tests have shown that the main antioxidants in *doenjang*, genistein, and omega-6 fatty acids are highly effective in preventing cancer.

Cardiovascular Risks and soybean

Cardiovascular disease, which occurs when an abnormality in the heart or blood vessels occludes the smooth flow of blood, includes high blood pressure, arteriosclerosis, hyperlipidemia, heart failure, and myocardial infarction. The isoflavones, fiber, saponins, lecithin, phytosterols, unsaturated fatty acids, and peptides in soybeans function to lower cholesterol and improve heart function. Because the chemical structure of saponins is similar to that of cholesterol, they hinder the absorption of cholesterol and even help expel cholesterol from the body. Bioactive peptides[1] serve to prevent a substance found in the blood called angiotensin, a compound that raises blood pressure. *Cheonggukjang* in particular contains a powerful protease (an enzyme that breaks down proteins) that acts as an excellent inhibitor to angiotensin-converting enzymes. The results of much research regarding the influence of soybean proteins on the level of lipids in the blood indicate that a daily intake of 70-80mg of the isoflavones contained in soybean protein (31-40g/day) significantly decreases the level of serum cholesterol in the body.

Thrombolytic (Clot-dissolving) Effect

Fibrin, a part of the fiber in our bodies, can adhere to our blood vessels and form fibrin clots, which typically happens when the body's blood vessels are damaged, so that after bleeding from a wound, the blood forms clots in the vessels and tissues. Symptoms of fibrin clots include spasms of the legs at night, heavy and stiff legs, and pain in the legs when standing or sitting for long periods. In normal circumstances the thrombolytic substance plasmin and its precursor, plasminogen, maintain an equilibrium between the formation and destruction of clots. In an abnormal situation, however, the flow of blood is restricted and therefore oxygen and nutrient cannot be supplied to specific tissues, thus triggering thrombosis (the formation of clots). The results of much research into the literature on this topic show that *cheonggukjang*, *natto*, *doenjang*, and other fermented soybean products, more than unfermented

1) Bioactive peptides are a specific array of amino acids containing chains of two to ten amino acids.

soybeans, excel in dissolving fibrin clots. Bacterial strains of bacillus in *doenjang* were found to have 3-4 times the thrombolytic effect than those strains found in *natto* and *cheonggukjang*, and *doenjang* made in the traditional way was 10 times more effective than *doenjang* that was mass produced.

Diabetes and Soybeans

Behind the stomach lies a small organ that plays an important role in digestion and immunity, the pancreas. The pancreas secretes insulin, which conveys blood sugar to each cell in the body. Diabetes occurs when complications with blood vessels arise, which can lead to damaged kidneys, heart, and peripheral nerves. Soybeans have a naturally low glycemic index, which means that their soluble fiber reduces the speed at which glucose enters the bloodstream and they sensitize cells to insulin, and thus they are effective in regulating blood sugar. The pinitol derived from in soybeans demonstrates insulin-like behavior by helping regulate insulin's signal system, thus having the effect of improving blood sugar. The trypsin and chymotrypsin (pancreatic digestive enzymes) in black soybeans especially promote insulin secretion by revitalizing pancreatic function and thus have the effect of bettering diabetes. The problem is that these soybean enzymes lose their vitality when heated (it has been shown that their effectiveness is limited at about 50℃), so in order to satisfy the expectation that the enzymes contained in black soybeans will be effective in reducing blood sugar, they need to be pickled in vineger or preserved by some other means. Outside of these enzymes, soybeans also contain blood sugar-reducing oligosaccharides.

Prevention of Aging and Soybeans

The entire time from birth until death can be called a process of aging, and one of the chief culprits in aging is free oxygen radicals. Our bodies manufacture an antioxidant called SOD (superoxide dismutase) as a defense system to block free oxygen radicals. The main role of SOD is to remove the toxicity of free radicals by using the electron transport system within the metabolic process to combine the excess free radicals with other substances before they oxidize, thus rendering them harmless. The various ingredients in soybeans that have antioxidant effects include isoflavones, anthocyanin and other phenol compounds, trypsin inhibitors, amino acids, peptides, phytic acid, and vitamins A, C, and E. Anthocyanins, part of the flavonoid family, are the red, blue, and purple pigments found in fruits and flowers, as well as the black pigments in black soybeans, which have been found to have four times the antioxidant activity of regular soybeans. Phytic acid has the ability to easily combine with metal ingredients like iron, so it prevents free radicals that would form when iron and oxygen combine from damaging cells. Fermented soy foods provide isoflavones that absorb easily in the body and increase the water-soluble browning substances that occur during the process of fermentation; they provide eight times as much antioxidant activity as unfermented soybeans.

Brain Function (Alzheimer's, Parkinson's) and Soybeans

The part of the human brain known as the cerebrum functions as the organ that controls all the body's activities and has a staggering 100 billion nerve cells, or neurons. Humans begin thinking from the time they are born. Neurons play a key role in the storage and usage of memory; as we age, the number of neurons in our cerebrum gradually decreases. Soybeans contain the phospholipid lecithin, which is effective in restoring the function of a neurotransmitter called choline. The activity of choline acetyltransferase, an element that decreases in those who suffer from Alzheimer's disease, increases with injections of estrogen; isoflavones function in a similar way to estrogen, and among the stimulative agents important for memory, isoflavones can substitute for estrogen.

Bone Health and Soybeans

New bone tissue can be formed in the body, and damaged bone tissue is dissolved by a complex process. Well-known bone diseases include osteoporosis and arthritis, which tend to appear as we get older. It has been reported that people who eat mostly animal-based protein have a higher rate of osteoporosis than people who eat mostly plant-based proteins. The calcium-saving effect of soybean proteins explains this phenomenon. Soybeans contain little of the sulfur-containing amino acid methionine, whereas the many sulfur ingredients contained in animal proteins prevent the reabsorption of calcium in the kidney, which is then washed away in the urine. Adding soy protein to the diet is an important way to prevent calcium loss via calcium excretion in the urine due to sulfuric amino acids. In clinical trials looking at the isoflavone content of soybeans, it was found that intake of soy not only increased bone density in post-menopausal women, but also suppressed unwanted angiogenesis, the cause of arthritis.

Weight Control and Soybeans

Obesity is correlated to many degenerative diseases and major disorders. For good health it is important to try to maintain a normal weight by reducing junk-food intake and making lifestyle changes. Trends moving in this direction increase every year. If we burn fewer calories than we ingest, or if the amount of calories we ingest and the amount we burn are balanced but something abnormal happens to our metabolism, we can become obese. The cure for obesity takes a rather long time and patience is required, but diet, exercise, and behavior modification together can make it work. Soybeans seem to be a great option for an optimum low-calorie, low-fat diet: the fiber gives one the feeling of being full while also preventing constipation, thus helping in weight control efforts, and the unsaturated fatty acids decrease neutral lipids, which is also effective in controlling body fat.

Chapter 06

The Future of Soybeans

1. Legumes and Nitrogen Fixation

The Discovery of Root-Nodule Bacteria

Long ago in the East, people knew that growing soybeans in the field enriched the soil. However, in the West where soybeans were introduced belatedly, people were not able to recognize the extremely important part soybeans played in farming until the discovery of root-nodule bacteria had been made. In 1885, M.W. Beijerinck, a soil microbiologist in the Netherlands, successfully separated leguminous bacteria responsible for nitrogen fixation. Leguminous bacteria are known to make the root tissue grow in lumps spread sparsely on the root, and are also called symbiotic free nitrogen fixation bacteria or root-nodule bacteria. Root-nodule bacteria

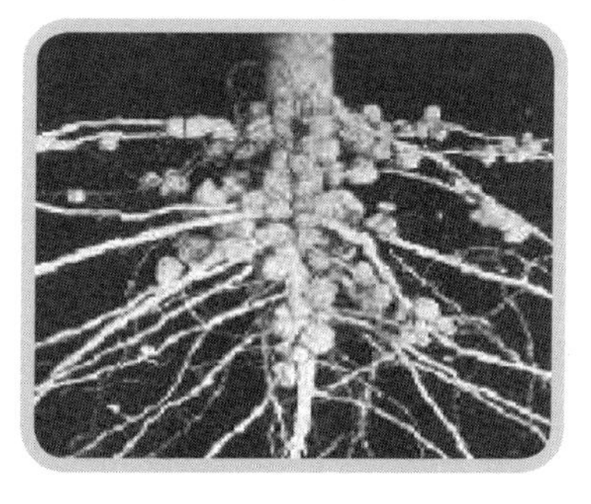

live symbiotically with higher plants, fixing free nitrogen, and the given bacterial strain differs depending on the type of plant. Most of plant species do not support root-nodule bacteria at all. The nitrogen content in the air is about 78%, but most plants cannot use nitrogen from the air. However, root-nodule bacteria living on the roots of soybeans are capable of converting nitrogen in the air into ammonia. Ammonia is converted into amino acids by means of enzymes, which then form proteins. These root-nodule bacteria turn the soybeans into an extraordinary plant whose protein content reaches about 40%.

Categorizing Legumes

Leguminosae are distributed all over the world, and about 13,000 species from 550 genera are found in the tropics, and in Korea 92 species from 36 genera are distributed around the country. Soybeans are an annual herbaceous plant, and their botanical name is *Glycine max* (L.) Merrill. When soybeans sprout, a single young root becomes the original taproot from which lateral roots sprout, and many root nodules occur on these roots. Leguminous bacteria in the root nodules fix nitrogen in the air so it can be used by plants.

These bacteria produce a nitrogen compound with the help of root-nodule bacteria living on the roots that can be utilized immediately by the plant. In other words, legumes function as a natural nitrogen factory. Genera and species of leguminosae can be told apart by looking at the shape of their seeds and seedlings. Seeds can be classified by observing the shape of the seed and the embryo bud's length, shape, height, and position. Also they can be classified by the bud's position, as in whether it is above or under the ground, or whether its first leaf has a leaf stalk or not, or by the shape of the bottom part of the leaf, and the shape and number of a seed leaves. Species having a seed leaf underground include pea, red bean, fava bean, pigeon pea and winged bean. Such a seed leaf does not have a hypocotyl growing at the bottom, which prevents it from sprouting up through the ground. Leaves of leguminosae are mostly compound. However, their first leaf is a single leaf, and compound leaves usually appear from the third leaf on.

Circulation of Nitrogen

Nitrogen is indispensable to all life, including animals, plants, and humans. People need at least 16g of nitrogen daily (composition of the air: nitrogen 78%, oxygen 21%, argon 0.93%, carbon dioxide 0.04%). However, nitrogen gas in the air does not react to biochemical events occurring in the body, though it comes in and out of the body during normal breathing. This is because nitrogen molecules are so firmly bound together that they begin to break apart only when heated at a temperature of 1,000℃ or above. Nitrogen fixation enables the plants to absorb nitrogen, converting it into protein. Animals, when they feed on the plants, can take in such protein. This then is expelled in animal waste in the form of ammonia or urea, returning to the soil or to the atmosphere through nitrogen separation. Legumes play a vital role in nitrogen circulation. Leguminous bacteria living on the legumes fix nitrogen in the air, enabling the legumes to use it, and whatever nitrogen remains in the soil can be utilized by the next crop.

Nitrogen Fixation

Nitrogen fixation refers to the conversion of nitrogen found in the air into ammonia. Nitrogen molecules in the air are joined with strong triple bonds and are therefore chemically close to inertness. This means that nitrogen in the air seldom reacts to or interacts with other chemicals. For this reason, nitrogen fixation is a process in which chemically inert nitrogen gas is converted into ammonia, a form of active nitrates. Nitrogen fixation occurs naturally, but it is also synthesized for industrial use. When nitrogen fixation is referred to, it usually indicates biological nitrogen fixation induced by microorganisms, though it can also occur due to physical causes such as lightning. Rhizobium bacteria living on the roots of soybeans fix 100kg to 400kg of nitrogen annually per hectare, which fertilizes soybean plants to the order of 43 bags of urea fertilizer.

The Symbiotic Relationship between Legumes and Root-Nodule Bacteria

The growth of nearly all plants is controlled by the nitrogen content in the soil. However, legumes are exceptional in that they become free from such restrictions by maintaining a symbiotic relationship with a soil bacterium called Rhizobium. Rhizobium is a genus of bacteria that fix nitrogen in the air, and when the bacteria enter into a host plant, they create nodules on the roots of that plant and live in the nodule.

When two different species living together share benefits and maintain a close relationship, it is called symbiosis. Plants provide the bacteria with a variety of nutrients indispensable to their survival while the bacteria supply fixed nitrogen to the plants. Soybeans are typical leguminous plants that utilize nitrogen in the air by fixing it through maintaining a symbiotic relationship with the root-nodule bacteria living on their roots.

2. Soybean Products Saving the Environment

Soybeans may be described in the single phrase, "a plant just like the mother of the earth." By fixing nitrogen in the soil, soybean plants grow into plants containing much protein; in turn they enrich the soil while supplying highly polymerized protein to humans. Natural circulation of nitrogen occurs in three different ways. First, nitrogen circulation is accomplished through the death of an animal or plant when they decompose, thus becoming a source of nitrogen; second, when nitrogen is fixed by means of lightning; and third, through the nitrogen fixation taking place in the root nodules of soybeans and leguminous plants. This was how nitrogen was circulated before artificial nitrogen was invented. Artificial nitrogen was developed in the early 20th century, opening the gate to the broad use of chemically produced nitrogen, which contributed greatly to supporting the exponential increases in population that took place over the span of about 100 years–but this use of chemical nitrogen was not without problems. Excessive use of chemical fertilizer induced eutrophication, or a worsening of water quality, which, in turn lowered productivity and increased the risk of oxygen deficiency in the water and the resultant mass death of the animals living therein. Moreover, production of chemical fertilizer is based on petroleum energy, the overuse of which increases the risk of oil depletion. The eco-friendliness of legume crops is garnering more attention than ever under these circumstances.

According to Dr. Hong Eun-hee, legume crops are capable of maintaining soil fertility and preserving the environment by fixing hundreds of kilograms of nitrogen per hectare annually. They are capable of fixing 3 tons of carbon dioxide and releasing 2.18 tons of oxygen, demonstrating their great atmosphere-purification capacity. Also growing large amounts of soybeans can reduce soil loss, effectively reducing the 50% soil loss of sloped land over 7%.[1)]

Environmentally Friendly Products on the Rise

Recently the consumption of soybean products and the range of their uses are increasing, people are awakening to the importance of environmentally friendly products. Soybeans offer seemingly unlimited possibilities for solving 21st-century environmental problems, including possibly as an industrial substitute for oil. Presently, soybeans are being used for a variety of purposes, and their range of application is expected to grow greatly. Soybean-based environmentally friendly products were first developed in the United States as oil prices began to rise sharply in the early 1970s.

Soy Ink

Since the soybean as a raw material is available at a moderate price and causes no environmental pollution in comparison with crude oil, this product is highly recommendable. Soy ink has many advantages over petroleum-based ink in that it uses recyclable raw materials; it provides vivid color; the recycling time of newspaper waste is greatly reduced due to the low content of volatile organic

1) Hong Eun-hi, History of soybean cultivation, Kong (Soybeans), Korea University Press, Seoul, 103-136(2005)

matter in the ink; a person's hands are not stained with ink when reading a newspaper; and ink is easily removed, making it easy to recycle printed paper.

Cosmetics Containing Isoflavones

Isoflavones, a functional substance of soybeans, are also used in cosmetics. Functions including anti-cancer effects, antioxidation, and suppression of free oxygen radicals and DNA aging remain intact in the cosmetics, promoting skin regeneration and anti-wrinkle effects.

Vegetable-based Detergent Made of Soybeans

This detergent's biodegradation degree reaches 99.5%. Its detergent function is so strong that the process of dirt removal is visible to the naked eye. It is produced by extracting protein from the soybean.

Soybean Fibers

Soybean fibers are created by making a dough of soy protein and putting it into a bath containing salt and acids for coagulation and making thread. Thread made this way is used to produce cloth, which may be dyed and processed to manufacture a garment. Soybean fibers are light and soft and therefore often called vegetable cashmere. They are also called environmentally friendly, healthful fibers. They are made of soybean protein, a major component of the soybean that is completely biodegradable. They are suitable as food packaging due to their stability, and the waste degrades rapidly in the soil at a speed much faster than any petrochemicals. Also, the waste is ground and recycled to produce protein-rich livestock feed.

Building Materials Made of Soybeans

Soybean building materials are made of 40% soy flour, 40% recycled newspapers, and 20% other materials. The result is like solid wood and resembles marble tiles. It is said to be suitable for high-quality flooring, cupboards, furniture, and decorative walls.

Bio-solvents

Bio-solvents are used for removing oil, adhesive, and paint. Soy-based bio-solvents are nontoxic and are used for removing contamination caused by petrochemical substances as well as soil contamination. Soy candles made of soy wax produce very little soot, and complete combustion takes place when they are lit, emitting no hazardous substance.

Biodiesel

Biodiesel refers to an environmentally friendly product containing oxygen that is made by allowing vegetable oil waste, including food-grade oil or soybean oil, to react with alcohol. It is nontoxic because it is edible vegetable oil, and more than 77% of it biodegrades in 28 days in the natural world. Biodiesel is expected to come into worldwide use because a reduction of 2.2 tons of carbon dioxide can be granted in return for biodiesel consumption of 1 ton according to the United Nations Framework Convention on Climate Change. The prospects of soybeans as a raw material for biodiesel production are very bright. Biodiesel, a diesel substitute, was made available in Korea beginning in May 2002 and has done very well there.

3. Hope Seen in Afghanistan

Producing 1 calorie of protein from beef by means of fossil fuels requires burning 54 calories of fossil fuels, while producing the same from soybeans requires burning only 2 calories of fossil fuels. This means that energy consumption by a person who ingests protein from soybeans is a mere 4% of the energy consumption of the person who ingests protein from meat, and the amount of carbon dioxide generated by the person obtaining protein from soybeans is also only 4% of that of the person who acquires protein from meat. Methane is second only to carbon dioxide among the gases that act as unstable factors affecting global climate.

Methane generates 24 times as many greenhouse gases as carbon dioxide and tends to raise its content in the air much more rapidly. According to the US Environmental Protection Agency, 25% of the methane gas in the atmosphere generated by human activity originated in livestock farming. They concluded that the environmental damage induced by producing one pound of beef was 20 times higher than that of producing the same amount of pasta. They said forging a new relationship between people and their food and the earth would constitute a historical revolution. It is apparent that environmental ecosystems and global food crises are the two most urgent issues for human existence. The worsening of ecological crises all over the world threatens the very existence of humans.

Brazil

"Livestock farming is the main culprit of the tropical rain forest devastation in Brazil. We are witnessing a scene of devastation where the tropical rain forest is being converted into a hamburger."[1] By the 2nd half of the 20th century, Brazil emerged as the largest beef exporter in the world and became the world's second largest soybean producer. The main culprits of Amazon rain forest devastation are the creation of large-scale ranches and soybean fields. In Brazil, soybeans are produced on large-scale farms developed by forest-clearing in mid-western regions such as Mato Grosso State. About 400 truckloads of 4-ton trucks are being loaded every day at Port of Santos on the east coast of Brazil alone. Soybeans mass produced in this way are exported as is, except for 20% of them, which are exported in the form of soybean oil and end up being used as livestock feed for cows. Soybean production in Brazil is growing sharply, as is the case in the United States. Amazon rain forests, which are called the lungs of the earth, are being devastated in order to grow soybeans. Growing soybeans may appear good at a glance because of increased food production. However, the fact of the matter is that Brazilian soybeans are directly transported to US livestock feed factories.

Afghanistan

Miracles are occurring in Afghanistan: from 2006 to 2008, 18 states began producing soybeans, 11 states produced them in 2009, and in 2010 all 34 states began producing soybeans. Dr. Steven Kwon, who worked for Nestle serving as Nutrition Director, visited Afghanistan one day and found many of the child-bearing women dying due to malnutrition. Also, Afghanistan's infant mortality rate

---

1) Norman Myers, *The Primary Source: Tropical Forests and Our Future*, 2nd ed., W.W. Norton & Company, New York(1992)

was the highest in the world. What occurred to Dr. Kwon as a possible solution were soybeans. He was confident that the protein and calcium in soybeans would improve the health of both mothers and infants. He began bringing soymilk to infants at a moderate price and continued as long as his resources permitted.

To his surprise, these infants recovered their health sooner than expected. He then began gradually increasing the amount of soybean milk while adding 10% soy flour to naan, their staple food, whenever it was prepared. Eventually he established Nutrition and Education International and began growing soybeans himself in Afghanistan. In just 7 years Afghanistan was turned from a country growing poppies into one growing soybeans in all 34 states.

4. Wild Soybeans vs. Genetically Modified (GM) Soybeans

Utilizing Wild Soybeans

Wild soybeans, which grow extensively in the fields and mountains of Korea, are the ancestral plants of cultivated soybeans. Cultivated soybeans have evolved from wild soybeans in Korea for a long time. Wild soybeans have a variety of useful characteristics, including functionality, resistance to specific diseases and insects, and particular cultivation features. Since soybeans comprise the main raw materials of traditional Korean foods such as *doenjang*, soy sauce, and bean sprouts, which boast a long history of consumption, it is imperative to secure an expanded base of self-sufficiency in soybean production, with particular emphases on environmentally friendly farming and food security. Recently, demand for Korean soybeans is rising due to consumers' distrust of imported

GM soybeans and emphasis on the stability and excellence of Korean soybeans (especially in terms of suitability for fermented soybean foods). Demand for highly functional soybeans with reinforced isoflavones and anthocyanins is also increasing. It is Korea's duty, as the country in which soybeans originated, to develop high value-added soybeans from the great variety of genetic resources available in her wild soybeans.

Rearing and Cultivating Examples of Functional Soybeans

Among the major food crops, soybeans are the only one for which Korea is the original home. 40% of the soybean seed is made up of protein, with fat occupying 20%; this nutritional composition offers a higher nutritional value than that of any other seed. The number of chromosomes in wild soybeans is identical to that of cultivated soybeans, and so hybridization between the two is possible. It takes a significant amount of research and continuous tests to develop a new variety of soybeans improved with a wild ancestor's useful characteristics and functionality.

Rearing and Cultivating a Variety with High Isoflavone Content

A new variety of soybeans called *agakong*, baby soybean, developed through interspecific hybridization, is highly functional with the highest level of isoflavone content known so far in the world. This *agakong* varietal is fortified with an isoflavone content of 10,000ug/g, which is six times higher than the *taegwangkong* or *cheongjakong* varietals, which are typical Korean soybean species.

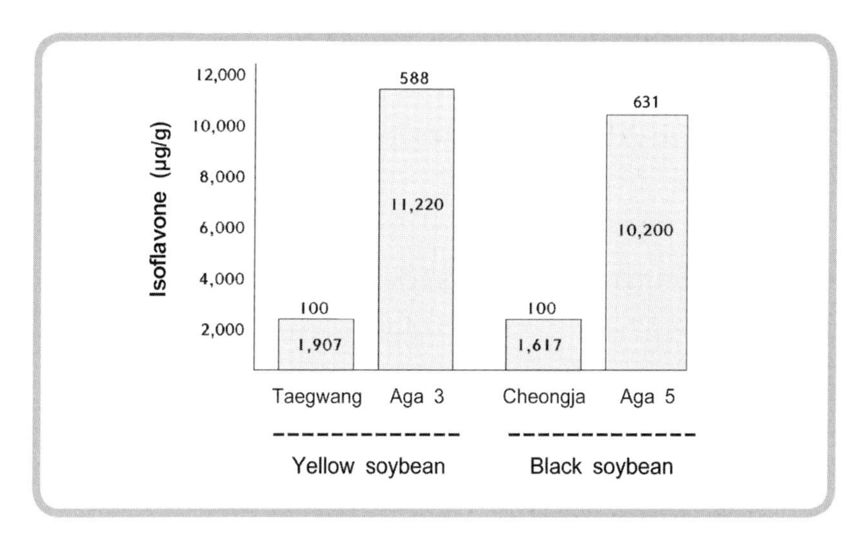

Comparison of the *Agakong's* Isoflavone Content

Rearing and Cultivating Soybeans for Livestock Feed

The most popular crop for livestock feed is corn. Although corn contains a high level of carbohydrates, its levels of protein and fat, which are needed by livestock, are low. However the mixed cropping of a new soybean variety, developed by crossing between a wild soybean and a common one, with a leafy corn variety can produce high nutritive cattle feeds whose protein and fat contents are 30 and 40%, respectively, higher than the corn feed alone.

Rearing and Cultivating Soybeans Containing a Higher Level of Oleic Acids

The functionality of olive oil to lower blood pressure is due to oleic acid (omega-9 fatty acids), whose content in olive oil reaches about 75%. In 2014 in Korea, a functional soybean whose oleic acid content is 4 times higher than that of regular soybeans was bred using traditional techniques and reared for cultivation. The ratio of omega-3 (linolenic acid) to omega-6 (linoleic acid) fatty acids in this new soybean variety is good, at 1:1. However, this ratio could be increased to 2:1 or 3:1 by means of cross-breeding with wild soybeans.

Fatty Acid Content and Ratio in the Species Containing a Higher Level of Oleic Acids

Soybean variety	oleic acid(%)	omega-3 : omega-6
Olives	75	-
Typical Soybean Variety	23	6~7:1
High Oleic Acid Soybean Variety (cultivars with cultivars)	80	1:1
High Oleic Acid Soybean Variety (cultivars with wild soybeans)	70	2~3:1

Developing and Utilizing GM Soybeans

Biotechnology-based genetic modification was first developed as an alternative option for dealing with food crises caused by continuous population growth and the resultant shortage of cultivation acreage, fortifying the functionality of food, developing energy resources, mitigating environmental problems and curing as-yet incurable diseases. GM (genetically modified) crops constitute new species developed using molecular breeding technology with a view to controlling vermin (insect resistance), weeds (weed killer resistance), and disease (virus and bacteria resistance); fortifying functionality (vitamins, etc.); and finding a way to increase food production. GM soybeans resistant to weed killers are a genetically modified variety bolstered with weed killer resistance, and these were successfully commercialized in 1996. These GM soybeans are not affected by herbicides, and for this reason it is very easy to kill weeds in a soybean field. GM crops are not allowed to grow unless their safety is ensured through a series of tests and evaluations. Importation of such crops is controlled by the importing country, which conducts a safety test before distribution. Among GM crops, the cultivation of soybeans has risen sharply, as evidenced by the drastic expansion of soybean cultivation acreage. This is because large-scale farming in the United States, Brazil, and Argentina necessitates the use of weed killers.[1)]

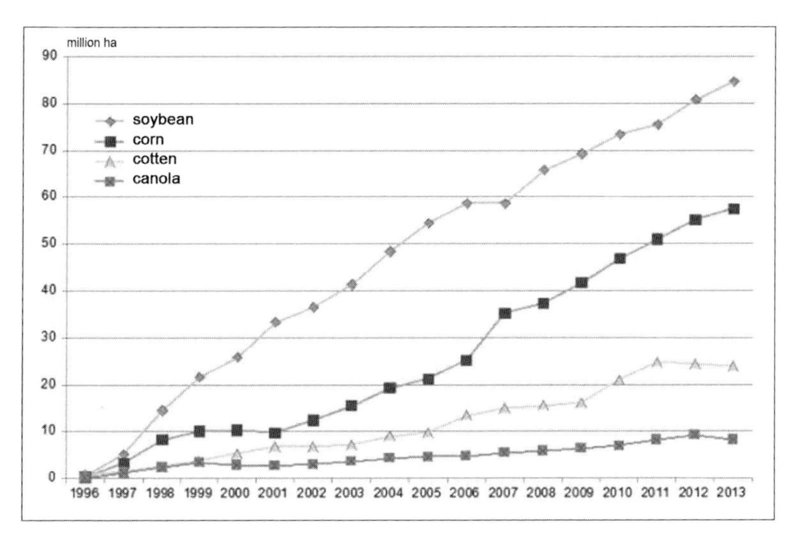

Trends of Change in Cultivation Acreage for Major GM Crops

Criticism against GMOs and the Labeling System

Denunciation of the safety of GMOs became serious as GM soybeans resistant to herbicides and GM corn resistant to vermin began to be commercially cultivated. Despite stubborn denunciation

1) Park Soo-chul, Kim Hae-yeong and Lee Cherl-ho, Understanding GMOs, Sikanyeon Publishing, Seoul(2015)

of GMO safety by environmental NGOs, the cultivation of GMO crops increased sharply in the major export countries such as the United States, Brazil, Argentina, and Australia, where crops are cultivated on a very large scale. According to 2014 statistics, GM crops are grown in 28 countries on a total acreage of 181.5 million hectares. GM soybeans comprise about 70% of the total soybeans cultivated throughout the world. Production of GM crops using biotechnology has now become a world trend, and the future of food problems is likely to be dependent on using this all-important biotechnology.

Meanwhile, the operation of the GMO labeling system differs from country to country. The United States, who takes the lead in developing and utilizing GMO crops, does not impose GMO labeling based on the principle of "substantial equivalence," which suggests regular crops and GM crops do not differ in composition. On the other hand, European countries, which need to protect their own agriculture, strictly regulate the distribution of GMO foods by mandating GMO labeling. In places like Korea, Japan, and Taiwan, where most of the food must be imported, GMO labeling is mandated, but food-grade oils, soy sauce, and glutinous starch syrup are exempt from labeling requirements because after processing, these products have no distinguishable foreign genes or proteins.

5. Production and Trade of Soybeans

The total global production of soybeans in 2013-14 was 285.3 million tons, and 112.83 million tons, or 39.5%, was exported to other countries. The five largest export countries of soybeans, Brazil, the United States, Argentina, Paraguay, and Canada, produce GM soybeans without exception, and these countries' GM soybean production (also known as adoption rate) stands at 93%, 94%, 100%, 97%, and 95%, respectively. The amount of soybeans exported by these countries has reached 107.53 million tons, occupying 95.1% of the total trade volume. Supposing that each country exported GM soybeans in proportion to their GM soybean adoption rate, the total GM soybean trade volume would be 101.08 million tons, occupying 90% of the total.[1)]

World Total Soybean Production, Major Producing, Export, and Import Countries (2013/2014)

unit: 1,000ton

Rank	Total Production: 285,302 Trade Volume: (39.5%) 112,829 (Export Rate Standard)					
	7 Production Nations		5 Export Nations		10 Import Nations	
	Nation	Production Amount	Nation	Export Amount	Nation	Import Amount
1	USA	91,389	Brazil*	46,829	China	70,364
2	Brazil	86,700	USA*	44,815	EU	12,950
3	Argentina	54,000	Argentina*	7,841	Mexico	3,700
4	China	12,200	Paraguay*	4,400	Japan	2,894
5	India	11,000	Canada*	3,471	Taiwan	2,350
6	paraguay	8,200			Indonesia	2,200
7	Canada	5,359			Thailand	1,798
8					Egypt	1,674
9					Turkey	1,608
10					Vietnam	1,350

*Major GM Soybean-producing
Source: USDA (2013-2014)

1) Park Soo-chul, Kim Hae-yeong and Lee Cherl-ho, Understanding GMOs, Sikanyeon Publishing, Seoul(2015)

Soybean Supply and Demand in Korea

Korea achieved a level close to soybean self-sufficiency up until 1970. The total soybean self-sufficiency rate was 86.1%, and the rate for edible soybeans reached 92.3%. The annual soybean consumption per capita was 5.3kg, which was not sufficient considering the amount of people suffering in absolute poverty at that time. Animal food consumption increased with economic growth, and this led to the promotion of livestock farming, which in turn sharply increased the demand for soybeans for livestock feed, causing the self-sufficiency rate to fall drastically, to 35% in 1980 and further to 20% in 1990. The self-sufficiency rate fell to under 10% as WTO trade liberalization set in following the Uruguay Round agreement.[1)]

Most imported soybeans are used for livestock feed. The total demand for soybeans in Korea reached 1.5 million tons in 2010, out of which 1.17 million tons (74%) were used for livestock feed, while the amount consumed for food was a mere 26% of the total soybean demand. Almost all local soybeans are being used for food, and shortages are filled with imported soybeans.

The domestic edible soybean consumption for 3 years beginning from 2008 averaged 0.406 million tons per year, while domestic production hit 0.126 million tons; the average self-sufficiency rate for edible soybeans, then, reached 31%. Considering the importance of soybeans, efforts should be made to achieve self-sufficiency in edible soybeans even in view of food security alone.

---

1) Lee Cherl-ho, Moon Huhn-pal, Kim Se-kwon, Lee Suk-jong, and Lee Ggotim, Food self-sufficiency - A condition for advanced country, Sikanyeon Publishing, Seoul(2014)

Soybean Supply, Demand and the Self-Sufficiency Rate in Korea

	1970	1980	1990	2000	2013
Production (1,000 tons)	232	216	233	113	123
Import (1,000 tons)	36	417	1,092	1,567	1.153
Food-grade Soybean Self-sufficiency Rate (%)	92.3	64.3	64.9	28.2	29.1
Soybean Self-sufficiency Rate (Including Feed) (%)	86.1	35.1	20.1	6.4	9.7
Annual Per Capita Consumption (kg)	5.3	8.0	8.3	8.5	8.0

Source: Ministry of Agriculture, Food, and Rural Affairs Statistics, 2014.

Import Status of Soybeans by Use

Soybeans used for livestock feed and for oil expression, such as for manufacturing soybean oil and hardened oil, are all imported GM soybeans. Soybean meal imported for manufacturing processed soybean products, hardened oil, and secondary processed-fat products used for pastries and bread making are all produced using GM soybeans, and they are classified as "miscellaneous" products. The import quantity of secondary processed-fat products, including soybean meal and hardened oil, is half the total soybean import quantity, and the import amount in 2012 was 1.539 million tons. Non-GM soybeans are used for producing food such as bean sprouts, tofu, and fermented soybean foods. The import quantity of soybeans used for oil expression and livestock feed amounted to 1.152 million tons in 2012, and the import quantity for non-GM edible soybeans was 0.319 million tons, occupying 28% of the total.

Local soybean production reached 0.129 million tons in 2012, and the supply amount of non-GM soybeans in Korea is estimated at 0.448 million tons.[1)]

In 2012 the amount of imported GM soybeans reached 730 million USD, and non-GM soybeans, 250 million USD. Imported non-GM soybeans occupied 22% of the total. It seems the reason for the lower importation percentage of non-GM soybeans stems from the fact that most of the soybeans imported for food were subject to tariff rate quotas.

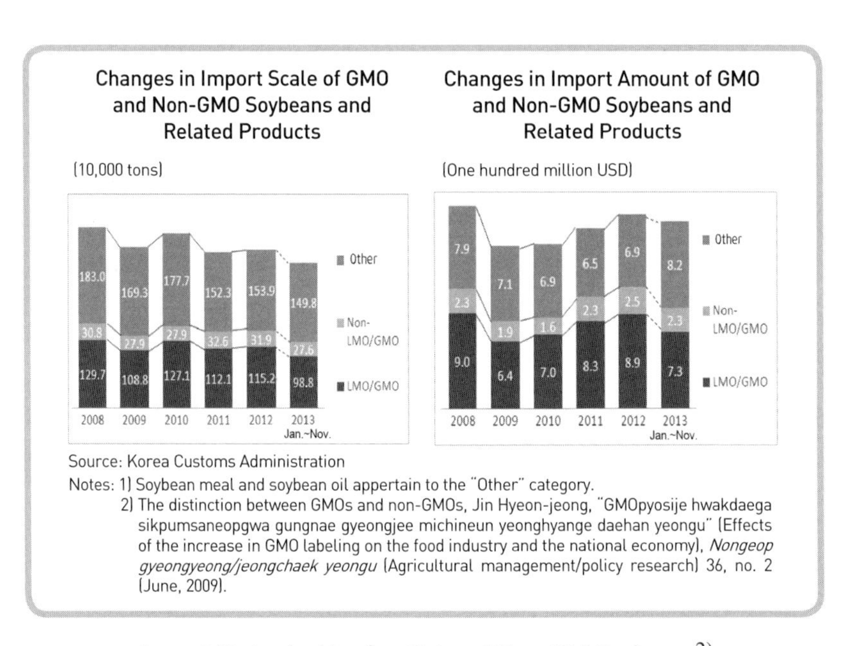

Source: Korea Customs Administration

Notes: 1) Soybean meal and soybean oil appertain to the "Other" category.

2) The distinction between GMOs and non-GMOs, Jin Hyeon-jeong, "GMOpyosije hwakdaega sikpumsaneopgwa gungnae gyeongjee michineun yeonghyange daehan yeongu" (Effects of the increase in GMO labeling on the food industry and the national economy), *Nongeop gyeongyeong/jeongchaek yeongu* (Agricultural management/policy research) 36, no. 2 (June, 2009).

Import Status by Use for GM and Non-GM Soybeans[2)]

---

1) Park Soo-chul, Kim Hae-yeong and Lee Cherl-ho, Understanding GMOs, Sikanyeon Publishing, Seoul(2015)

2) Lee Bu-hyeong, et al., A study on the prospects of environmental change in the supply and demand of major crops at home and abroad, Hyundai Research Institute, Seoul(2014)

6. The Future of Korean Soybeans

In prehistoric times, soybeans were first used for food by people of the Eastern (*Dongyi*) tribes, who stood their ground on the Korean peninsula and in southern Manchuria. Since then, soybeans have become a staple food for people living throughout the world, and the prospects are bright in that they can be used not only for functional food and drugs but also for major industrial materials, thanks to their nutritional functionality and physical properties. According to the dietary guidelines for Americans it is recommended that people should take in more than 25g of soybean protein per day to prevent various metabolic and adult onset diseases.

Soybeans are a great inheritance of the Korean people. Not only are fermented soybean products basic to the taste of much Korean food, but they also serve as indicators in global history that demarcate the Northeast Asian cultural sphere. The Korean diet of *doenjang* (fermented soybean paste) and *cheonggukjang* (fast-fermented bean paste) stew, which is boiled in rough earthenware vessels and served at modern dining tables, is an example of the unique, 8,000-year-old food culture showing its distinctive colors.

Nutritionally, soybeans have reached imperative status on the Korean dining room table. Koreans will never miss a nutritionally balanced, tasty meal, provided that cooked rice (*bap*) and soybeans (in the form of *doenjang* stew, soybean sprouts, or tofu) are served. Korean soybean-derived foods constitute an important subject in the framing of the origin and development process of Korean food culture.

From the viewpoint of nutrition anthropology, the fact that the Eastern Yi (Archers) tribes cultivated soybeans for food played a vital role in their ruling over the region not only culturally but also militarily. In fact, the Eastern Yi tribes were able to compete with the Chinese, as seen in the ancient literary histories of China, and they played a leading role in the early years of Chinese civilization.

A book titled *Ziyuan* (Character origins) quotes a late Tang dynasty history text depicting the Eastern Yi tribes as natives who had lived first in the region located in and around the central and lower reaches of the Yellow River; it further depicts them as people of gentle moral character with a duty-bound culture. Some scholars of ancient history indicate that the Eastern Yi tribes were the native founders of the Shang dynasty. In the history book *Hanshudilizhi*, the people of the Eastern Yi tribes were depicted as gentle and different from neighboring barbarians. Such a description was exceptional as the Chinese people used to look down on other races, calling them barbarians. To the Korean people, the ancestors of whom once emerged as a leading power in Northeast Asia, soybeans continue to be a staple food, second only to rice, and therefore efforts must be made to achieve self-sufficiency in this basic foodstuff.

To Koreans, soybeans are a silent reminder of our forgotten ancient history. It is logical in that sense to say that the Korean Soyworld Science Museum founded in Yeoungju city is more than merely a place for exhibiting the soybean; it is a history textbook displaying the Korean people's vision of the world.

Korean Vision of Unification for Soybean Self-sufficiency

Achieving self-sufficiency in edible soybeans is a national task that must be fulfilled to achieve food security. Demand for edible soybeans is estimated at 625,000 tons on the Korean peninsula, where 70 million people will live after unification.[1] North Korea is a prime candidate for cultivating soybeans as it has plenty of mountainous areas and large fields. Soybean cultivation acreage in North and South Korea reached 75,000 and 90,000 hectares, respectively, as of 2009. The yield potential of soybeans in South Korea is 1.65 ton/ha, to North Korea's 1.16 ton/ha, which is about 70% of South Korea's yield potential. By converting North Korea's corn fields into soybean fields and then raising the yield potential to the level of South Korea, a total of 0.625 million tons of soybeans could be produced in the combined fields of 0.375 million hectares.

Edible Soybean Self-Sufficiency Plan after Unification

	South Korea	North Korea	Total
Food-grade Soybeans Total Demand (2010, 1,000 tons)	438	187	625
Food-grade Soybeans Total Production (2010, 1,000 tons)	139	154	293
Soybean Cultivation Area (2010, 1,000ha)	71	90	161
Post-unification Cultivation Area (1,000ha)	80	300*	380
Post-unification Food-grade Soybeans Yield (1,000ha)	132	495**	627

*Based on 210,000ha of North Korea's 500,000ha of corn fields being converted to soybean fields.

**Based on North Korea's soybean yield capability (1.16 tons/ha) increasing to South Korea's level (1.65 tons/ha).

1) Lee Cherl-ho, Moon Huhn-pal, Kim Se-kwon, Park Tae-Kyun, Kwon Ik-boo, *Korea Unification and Food Security,* Sikanyeon Publishing, Seoul(2015).

콩 스토리텔링

한국콩박물관건립추진위원회 편

KFSRF 한국식량안보연구재단 도서출판 식안연

콩 스토리텔링

초판 1쇄 발행 2017년 4월 5일
2쇄 발행 2017년 5월 18일
3쇄 발행 2017년 10월 27일

발행인 이철호(한국식량안보연구재단)
발행처 도서출판 **식안연**
주 소 서울시 성북구 안암로 145, 고려대학교 생명과학관(동관) 109A호
전 화 02-929-2751 / 팩 스 02-927-5201
이메일 foodsecurity@foodsecurity.or.kr
홈페이지 www.foodsecurity.or.kr
편집·인쇄 한림원(주) http://www.hanrimwon.com

ISBN • 979-11-86396-36-0 **정가** 20,000**원**

국립중앙도서관 출판예정도서목록(CIP)
콩 스토리텔링 / 한국콩박물관건립추진위원회 편. -- 서울 : 식안연, 2017
한영대역본
ISBN 979-11-86396-36-0 93400 : ₩20,000
콩[豆]
524.42-KDC6 / 633.34-DDC23 CIP2017007421

콩 스토리텔링

머리말

한국은 콩의 재배와 이용을 시작한 콩의 종주국임에도 불구하고 대부분의 한국 고대사가 그랬듯이 중국 문화에 묻혀 세계에 제대로 알려져 있지 않다. 한국식품연구원 초대원장을 지낸 권태완 박사님은 2001년 한국콩박물관건립추진위원회를 결성하고 콩의 역사와 이용을 제대로 알리는 박물관을 건립하는 일을 시작하였다. 아래는 추진위원회를 시작하면서 발표한 취지문이다.

한국콩박물관건립 추진 취지문

우리가 인류 역사상 제일 먼저 콩을 심고 먹기 시작한 민족이라는 사실이 그 동안의 연구를 통해 밝혀지고 있다. 동북아(東北亞)에서 콩의 식용(食用)은 3천년 이전으로 거슬러 올라가는데, 중국 남부와 동남아, 그리고 일본에는 기원전 3세기부터 전파(傳播)되었고, 18세기가 되서 유럽에 소개되었으며, 세계 제2차 대전을 거치면서 미국의 경제작물로 재배되어 지금 전 세계적으로 확대되고 있다.

한편, 콩이 쌀을 영양적으로 완벽하게 보완(補完)한다는 것은 이미 알려진 바이나, 최근 암과 심장병, 그리고 골다공증 등 성인병(成人病)을 예방하고 치유(治癒)할 수 있다는 사실이 밝혀지고 있다. 이렇게 콩은 곡식을 주식으로 하는 민족의 귀중한 식량자원으로서 그들의 삶을 오늘날까지 뒷받침해 왔지만, 기름이나 사료작물로 쓰던 구미(歐美)에서도 최근 식품학적 진가(眞價)를 알게되면서 콩의 직접 식용이 매우 빠른 속도로 확산되고 있다. 이와 같이 동북아 기원의 콩은 그 생산이나 다양한 용도로 볼 때, 이제 세계의 작물, 세계인의 먹거리가 된 것이다.

우리 조상의 옛 터전이던 만주와 한반도에서 기원하여, 기나긴 세월을 통하여 우리 민족의 영양과 건강을 보살피고 또한 삶을 함께 해온 이 콩이 바야흐로 세계적

무대에 등장하고 있다. 21세기는 문화(文化)의 세기(世紀) 즉 문화가 나라의 양심(良心)과 역량(力量)을 가늠하는 시대이다. 이 지구상에는 수많은 박물관이 저마다 문화의 상징으로 여기 저기 세워져 있으나, 아직 콩 박물관은 한 군데도 없다. 이것이야말로 우리가 콩 문화의 종주국(宗主國)으로서 콩 전문박물관을 세우도록 하늘이 점지(點指)하신 일이 아니겠는가! 콩 문화와 이용기술의 중심이 될 세계적 차원의 박물관을 건립하여 우리 콩 문화를 지구촌에 널리 알리고, 콩 식품을 발전 전파함으로서 인류건강에 이바지 할 때이다.

이 박물관은 단순히 콩에 관한 옛것을 모아 보관 전시하는 수준을 뛰어넘어 전 세계의 콩 관련 자료와 문헌(文獻)을 수집 분석하고 연구하며, 국제적인 정보교류와 연구 교육의 중심이 되는 미래지향적이고 창조적인 콩 문화 과학 박물관이 되어야 할 것이다. 따라서 전천후 콩 재배온실, 콩 식품제조 및 가공실습실, 콩 전문 음식점 등을 병설(倂設)하여 콩에 관한 모든 것을 한자리에서 보고 배우고 생각하며 먹고 체험할 수 있도록 할 것이다. 이렇게 이 박물관은 자립형으로 운영 발전 될 것이다.

오늘 이 시점에서 우리가 콩 박물관을 세우고자함은 암울했던 지난 20세기에 잃어버린 천년(千年)의 우리 문화를 되찾아 2천년 대를 살아갈 우리 후손(後孫)에게 선조의 슬기와 창조적 정신을 물려 주려함에 있다. 이 일을 위하여 우리 모두의 역량을 모아야 할 것이다.

2001년 9월

한국콩박물관 건립 추진위원회

발기인 : 권태완 · 김석동 · 김석민 · 김준영 · 유용환 · 이철호

이영택 · 장학길 · 정장섭 · 정재원 · 조세영 · 홍은희

이 사업은 1998년 대산농촌문화재단의 연구비 지원으로 수행된 권태완, 권신한, 이철호, 홍은희 공저의 '국제규모의 콩박물관 건립에 관한 타당성 조사연구'를 기초로 하여 역사적 고증과 과학적 발전을 망라하는 자료수집으로 시작되었다. 2005년 위원회는 국내 콩 관련 주요 연구자 대부분이

참여하는 ‘콩, 大豆, Soybean (15장 794쪽, 고려대학교출판부)’ 책자를 출판하였다. 이 책은 콩의 이용역사를 비롯하여 콩과 관련된 고고학적 유물, 유적, 야생콩의 분포와 재배 육종 역사, 콩의 성분과 기능성, 콩을 이용한 음식과 식단, 콩의 산업적 이용과 생산 유통 현황과 전망까지 국내외의 자료를 종합적으로 수렴한 참고서이다. 책의 목차와 집필자를 소개하면 아래와 같다.

[머리말(권태완), 1. 콩의 이용역사(이철호, 권태완), 2. 선사고대유적의 콩(조현종), 3. 장류문화와 토기(신숙정), 4. 콩 재배역사(홍은희), 5. 콩 품종과 육종(김석동, 이영호), 6. 콩의 가공특성(김우정), 7. 두유 두부의 제조역사와 현황(손헌수), 8. 콩 발효음식(신동화, 이효지), 9. 콩 발효식품의 건강기능성(박건영), 10. 우리나라 콩음식들(이효지), 11. 다른 나라의 콩 이용 음식(조정순), 12. 콩음식의 영양가와 기능성(승정자), 13. 콩의 산업적 이용(지규만), 14. 콩기름과 그 부산물(이경일), 15. 콩의 생산 및 유통 현황과 전망(조세영)]

이 책에 근거하여 위원회는 2008년 사이버 콩박물관(www.soyworld.org)을 개설하였다.

2011년 위원회는 이 책에 근거하여 콩 스토리텔링 자료를 만들기로 하고 유미경 위원이 이 일을 맡아 수행하였다. ‘콩 스토리텔링’은 2015년 경상북도 영주시에 건립된 콩세계과학박물관 전시 내용에 중요하게 사용되었으며 박물관 도록에 수록되었다. 도록의 기획은 이철호, 글 유미경, 감수에는 김석동, 송희섭, 이영호, 문갑순, 황영현 위원이 수고하였다.

2015년 이 책을 영문화 하기 위해 미국 하버드대학 대학원에서 한국문학을 전공한 다이아나 에반스(Diana Evans)씨에게 번역을 요청하여 1년여에 걸친 열정적인 작업으로 세계에 내놓게 되었다. 국영문 합본의 감수와 교정은 김석동, 송희섭, 이영호, 문갑순, 황영현, 황인경 위원이

수고하였다.

권태완 박사님을 비롯하여 이 모든 일에 헌신적으로 협력하신 모든 분들께 진심으로 감사와 경의를 표한다. 이 일을 통해 우리세대가 반드시 감당해야 할 역사적 소명을 미약하나마 시작할 수 있었던 것에 대해 감사드린다.

2017년 4월

한국콩박물관건립추진위원회 제3기 위원장

한국식량안보연구재단 이사장 이철호

목 차

Chapter 01

콩의 기원

1. 우리는 언제부터 콩을 먹었을까

인간의 역사는 먹을거리를 얻기 위한 노력으로 시작되어 오늘에 이르고 있다고 해도 과언이 아닐 정도로 식량 확보는 인간의 생존에 필수적이었다. 그래서 지역마다 주어진 환경에서 독창적인 음식과 문화가 발전되어 왔다.

고고학 자료에 의하면 한반도의 원시토기문화시대(BC 6000~3000)에 이미 누룩의 제조기술을 비롯한 술 양조기술, 김치와 젓갈 발효기술들이 상당한 수준으로 발달하였을 것으로 보고 있다. 세계적으로 보면 야생콩은 한국, 대만, 일본, 중국 양자강 연안일대, 만주, 시베리아 등지에 분포되어 있다.

우리나라에서 야생콩은 제주도에서부터 함경북도에 이르기까지 골고루 분포되어 있는데 주로 산기슭, 초지, 덤불이 우거진 곳에 자생하고 있다. 한반도 곳곳에는 신석기, 구석기, 청동기, 철기시대를 망라해 탄화콩의 유적과 유물들이 발견되고 있다. 한반도 후기

구석기 시대 유적(약 4만년전)으로 알려진 제천 점말용굴에서 콩과 식물 화분(花粉)이 발견된 기록도 있다.[1)]

원시토기문화와 콩

콩은 날것으로 먹으면 심한 설사를 일으킨다. 콩에는 단백질의 소화를 막는 트립신 인히비터(trypsin inhibitor)가 들어 있기 때문이다. 석기시대 사람들은 콩을 먹을 수 없는 독초로 여겼을 것이다. 콩을 식량으로 만든 계기는 기원전 6000년 전후에 대한해협을 중심으로 발달한 원시토기문화의 영향이다. 원시토기의 광범위한 이용의 흔적은 한반도와 일본열도 남부를 연결하는 해안지역에 이르며 대한해협(Korea Strait) 연안이 현재까지 발굴된 바로는 세계에서 가장 앞서고 있으며, 따라서 이 지역을 원시토기문화의 발상지로 보고 있다.[2)]

해변가의 채집인으로 살던 한반도 원주민들이 토기를 만들어 인류 최초로 물을 담아 끓이는 기술을 개발했고 콩을 끓여먹으면 설사하지 않고 먹을 수 있다는 것을 발견한 것이다. 토기에 바닷물을 담고 주변에서 채집한 수산물과 채소를 넣고 끓이면 찌개가 된다. 이 과정에서 필연적으로 바닷물에서 소금을 만드는 방법을 알게 되고 소금으로 염장 발효하는 기술이 개발된다. 젖은 곡물과 뿌리를 토기에 보관하면 곰팡이가 자라 누룩이 되고 술이 된다. 토기를 사용함으로서 바닷물에 푸성귀를 담아두면 유산균이 자라 김치가 되고 생선을 담아두면 젓갈이 된다. 이와 같이 대한해협

1) 이철호, 한반도와 동북아시아 구석기시대 식생활 환경, 민족문화연구제31호, 415-458(1998)

2) Barnes, G.L., The rise of civilization in East Asia, The archaeology of China, Korea and Japan, Thames and Hudson Ltd., London(1993)

주변에서 시작된 원시토기문화는 끓임(boiling)문화와 발효문화의 근원이 되는 것이다.[1)]

콩과 동이족

동이족(東夷族)은 중국의 고문헌에서 한반도를 포함한 동북아시아 지역에 사는 민족을 지칭하는 말이다. 「설문해자說文解字 AD 100」에 의하면 이(夷)는 大(큰대) 弓(활궁), 人(사람인)의 합자(合字)로 큰 활(대궁)을 사용하는 부족 명칭에서 유래하였다. 이성우(1990)박사는 "인류사상 콩을 가장 먼저 음식으로 사용한 민족은 동이족 중에서도 예맥족(濊貊族)"이라고 결론짓고 '동북아 국가형성기(BC 4000~2000)초엽'이라고 하였다.[2)] 그는 북부 유목민들이 백두산을 중심으로 한 남만주와 한반도에서 농경정착을 시작한 신석기시대에 콩의 경작이 시작되었고 초기 청동기시대(BC 1500)에는 한반도를 비롯한 동북아시아에서 콩의 식용이 보편화되었을 것으로 보고 있다.

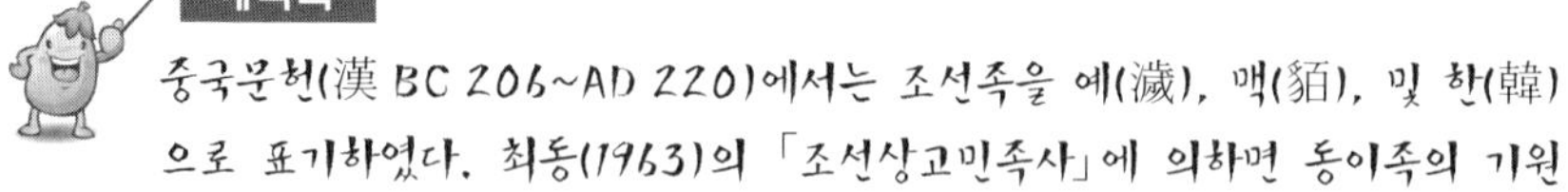

예맥족

중국문헌(漢 BC 206~AD 220)에서는 조선족을 예(濊), 맥(貊), 및 한(韓)으로 표기하였다. 최동(1963)의 「조선상고민족사」에 의하면 동이족의 기원은 기원전 3천년경 중앙아시아 바빌로니아 문화권 민족의 일부가 시베리아 남방으로 동진하여 송화강 연안에 정착한 후 원주민과 혼혈하여 이룬 민족이다. 이들이 남진하여 세 갈래로 나누어지게 되는데 중국 하북 지방, 요하 연안, 그리고 조선반도에 정착한 동이족이라고 하였다. 진시황의 만리장성 축조(BC 225) 이후에는 하북과 산동지방의 동이족은 한족에 완전히 흡수되고 남만주와 조선반도에 걸친 동북아 거류 민족만을 동이족(東夷族)이라 부르게 된다.

---

1) 이철호, 동북아시아 원시토기문화시대의 특징과 식품사적 중요성, 민족문화연구 제32호, 325-357(1999)

2) 이성우, 고대 동아시아 속의 두장에 관한 발상과 교류에 관한 연구, 한국식문화학회지, 5(3), 313-316(1990)

중국에서는 일찍이 콩을 일컫는 말로 대두 이전에 융숙(戎菽)이 있었다. 이는 콩이 원래 중국 민족의 작물이 아니라 만리장성 너머 그들이 융(戎)이라고 일컬었던 동이족의 작물이었음을 간접적으로 시인하는 대목이 아닐 수 없다. 반면 중국 남부와 동남아, 그리고 일본 등으로의 전파는 기원전 3~4세기경으로 보고 있다. 최근 발굴된 충북 옥천 대천리 신석기시대 집터에서는 쌀알과 콩류 등 20여점의 탄화곡물이 발굴되었다. 방사성 탄소연대 측정법으로 분석한 결과, 신석기 후기인 기원전 3000~3500년 사이로 판명되어 한반도에서 가장 오래된 콩의 출토 사례로 거론되고 있다. 그동안 나왔던 신석기 곡물 유물들은 주거지와 관계없는 토탄층에서 나온 것들이어서 야생종인지 아니면 농경용인지 가늠하기 어려웠다. 콩을 식용해 온 역사가 곧 우리나라 역사의 시원(始原)과도 맥을 같이 하는 5천년의 역사라는 점은 굉장히 시사적이고 흥미로운 일이다.

2. 콩의 원산지를 찾아서

식품을 구입할 때 보면 원산지 프리미엄이라는 게 있다. 비슷한 제품이라도 원산지 가공품이라면 가격이 더 비싸도 잘 팔린다. 원산지 논쟁을 차치하고라도 우리 민족은 수천 년 전부터 콩을 이용해 여러 콩 가공식품을 만들어 먹어왔다. 북 애리조나 대학 환경유지센터 소장인 게리 나브한(Gary Nabban)은 "우리는 우리 조상이 먹고 마신 결과물"이라며 "만약 우리 조상이 한 지역에 오래 살았다면 우리가 이 환경의 음식들에 유전적으로 적응되었을 가능성이 높다"고 하였다.[1] 그런 차원에서 보면 우리의 유전자 속에 콩과 장(醬)의 DNA가 흐른다는 표현은 단순히 수사적인 차원을 넘어서는 일이다.

---

1) 게리 나브한, 지상의 모든 음식은 어디에서 오는가? 아카이브(2010)

콩의 원산지를 밝히기 위해서는 콩의 재배기원에 대한 문헌적인 연구결과, 작물학계의 입장, 고고학적인 증거, 콩 문화의 현주소 등을 모두 검토해야 한다. 만주와 한반도를 콩의 원산지로 보는 우리 측의 주장은 일본의 후쿠다(福田) 박사가 주창한 '야생콩의 분화가 많이 된 곳이 원산지'라는 주장을 토대로 하고 있다.[1] '중국 원산지설'을 주장하고 있는 대표적인 학자는 미국의 하이모위츠(Hymowitz) 박사인데 그 주장의 근거는 기원전 11~6세기까지의 시(詩)를 모은 중국의 고전 「시경詩經」의 내용이다. 이 책에 콩을 의미하는 숙(菽) 자가 처음 기록되어 있다는 것이다. 기원전 11세기라면, 중국의 동북부는 한족이 아니라 동이족이 거주하고 있는 땅이었다. 과거 우리 민족의 활동무대였던 중국의 동북부, 만주 일대는 세계적으로 콩의 원산지로 널리 인정받고 있는 곳이다. 하지만 그에 못지않게 지금 우리가 터전으로 삼고 있는 한반도도 콩의 야생형인 돌콩이 널리 자생하고 있고, 재배종의 유전적 다양성이 매우 높아 콩의 재배기원지 중의 하나로 여겨지고 있다.

권신한 박사는 '만주일대에서 기원전 2500년경부터 재배되던 콩이 한반도로 전파되어 농작물로 재배되기 시작한 것은 기원전 2000~1500년경일 것'으로 추정하였다. 그는 또 "한국에서 재배되는 재래종은 초장, 숙기, 엽형, 종피색, 종실 크기, 지방 및 단백질 함량에 있어 현재까지 세계 각국에서 보고되어 있는 콩의 각종 특성을 모두 보유하고 있다. 이는 우리나라에서 재배되고 있는 조상 전래의 계통에 수많은 변이가 축적되어 있음을 증명하는 것"이라고 한 바 있다.[2]

1) Fukuda, Y., Cytogenetical studies on the wild and cultivated Manchurian soybeans (Glycine, L.), Japanese J. of Botany (1933)

2) 권신한, 대두의 기원, 한국콩연구회지, 2권, 4-8(1985)

문헌연구

「일주서」는 기원전 6세기 무렵에 나왔는데, "산융(山戎)은 동북이이(東北異夷)다. 융숙(戎菽)이 나는데 큰 콩(巨豆)이다"라는 내용이 있다. 사마천의 「사기史記」에서는 "BC 623년에 산융이 연나라를 쳤는데, 연이 위급을 제나라에 알리자 제나라의 환공이 연을 구해주고, 북으로 산융을 정벌하고 고죽국 지역까지 갔다가 융숙을 얻어 돌아왔다. 제환공은 이 융숙을 이웃 노나라에 주었다."고 하였다. 「관자管子」에서는 "제나라의 환공이 북쪽으로 산융을 쳐서 겨울파와 융숙을 가져와 온 세상에 펼쳤다(北伐山戎, 出冬蔥與戎菽, 布之天下)"고 하였다. 위 3종의 역사서를 참고해보면 '산융의 콩'이 콩의 역사에 있어 '최초의 X'가 아닌가한다. 최덕경 부산대 교수에 의하면 "중국의 제반 역사 기록들을 참조하면 대두는 주나라 초기 중국 동북지방에서 재배되기 시작하여 춘추 중기 이후 화북에 보급되었으며, 진한 이후 중국 전역으로 재배지역이 확대되면서 '숙(菽)'의 명칭이 '대두(大豆)'로 바뀌었다"고 하였다.1)

최초의 X

문화인류학자인 제레드 다이아몬드 「총·균·쇠」에 따르면 유럽 최초의 인류화석이든 멕시코에서 작물화된 최초의 옥수수이든, '최초의 X'가 존재한다. 「일주서」, 「사기」, 「관자」에서 말하는 산융에 대한 언급, 또 여러 학자들의 자료를 참고하여 대릉하 유역의 조양 일대를 콩의 재배기원에 있어 '최초의 X'로 보고자 한다. 즉 현재의 조양이 옛 산융이라는 것이다. [삼국유사]에서 말하는 고조선 최초의 수도는 아사달이고, 이를 한문으로 옮기면 '조양(朝陽)'이다. 조양 일대는 고조선의 대표 유물인 비파형 동검이 발굴되는 곳이기도 하다.

1) 최덕경, 대두재배의 기원론과 한반도, 중국사연구, 제31집, 중국사연구회(2004)

콩 전파의 역사

일반적으로 서기 700년까지는 중국 남부와 동남아 전역에 콩이 전파되었다고 보고 있다. 동남아로 콩이 전파된 것은 중국 화교들의 동남아 이주 역사와 밀접한 관계가 있을 것으로 판단된다. 이미 4세기부터 중국의 가난한 사람들이 동남아에서 남천민족(南遷民族)을 이루게 되었지만 9~13세기, 많은 사람들이 대규모로 화남으로 이주하게 되면서 콩 재배와 이용방법이 인도차이나 반도를 비롯한 동남아지역에 본격적으로 알려졌을 것으로 추측된다. 콩이 유럽에 알려진 시점은 1712년으로 일본에 다녀온 독일 학자 캠퍼(Engelbert Kaempfer)가 개인적인 호기심으로 콩을 소개하였다는 것이고, 공식적으로는 1739년 프랑스 선교사가 중국으로부터 콩 종자를 가져와 파리식물원에서 재배한 것이 최초이다. 1790년 영국 식물원에서 콩이 재배된 이래 당시 영국의 식민지였던 동서 아프리카에서도 재배실험이 이루어졌다.

콩이 미국에 전파된 경로는 두 갈래다. 하나는 1764년 동인도 회사 선원이었던 사무엘 보웬(Samuel Bowen)이 중국 광동에서 살다가 조지아주 사바나로 와서 콩을 재배했다는 것이고, 또 하나는 당시 프랑스 대사였던 벤자민 플랭클린(Benjamin Franklin)이 1770년 영국에서 콩씨를 구해 집(필라델피아)에 보냈다는 편지로 알 수 있는 것이다. 18세기 후반 동양의 콩이 서양에 알려지기 시작했지만, 경제적 작물로 관심을 끌게 된 것은 그로부터 1세기가 지난 후의 일이었다. 아편전쟁(1840~1842) 이후 미국의 농학자들은 중국인들이 콩을 식용으로 이용하는 것을 보고 '들판의 젖소'라고 칭하고 그 재배기술을 본격적으로 연구하기 시작하였다. 20세기 들어 제1차 세계대전과 2차 세계대전을 겪으면서 콩은 사료용, 녹비작물에서 '신데렐라

작물’ ‘기적의 작물’로 불리며 세계의 주요곡물이 되었다.[1)]

현재 세계 제1위 콩생산국인 미국은 20세기 초 한중일 3국에서 많은 콩 유전자원을 수집해간 일이 있다. 1929~31년 미농무부에서 파견된 팔먼 도셋(Palmon Dosett)과 윌리엄 모스(William Morse) 등은 총 4,578점의 콩 유전자원을 수집하였는데, 한국에서 3,379점(74%), 중국 동북부에서 622점(14%), 일본에서 577점(13%)이었다. 이를 보면 한반도에 얼마나 다양한 콩 유전자원이 있었는지 짐작해 볼 수 있다.[2)]

콩 전파의 역사

- 만주 일대 콩의 재배(BC 25세기)
- 한반도 일대 콩의 재배(BC 20~15세기)
- 중국 남부(BC 4세기)
- 일본(BC 3세기)
- 동남아에서의 재배(800년대)
- 캠퍼, 독일에 콩 소개 기록(1712)
- 파리식물원에서 콩 시험재배(1739)
- 사무엘보웬, 미국 조지아주 사반나에 콩 심음(1765)
- 벤자민 플랭클린, 미국 필라델피아 바트람 식물원에 콩 보냄(1770)
- 독일식물원에서 콩 시험재배(1786)
- 런던식물원에서 콩 시험재배(1790)
- 페리제독, 일본에서 콩 종자를 갖고 감(1852)
- 미국 농무성 주관으로 7개국으로부터 65종의 품종도입(1898)
- 윌리암 모스의 <동양의 식물탐험대> 콩 종자수집(1929~1931)
- 미국, 세계 콩 생산량의 50% 점유(1935)
- 브라질, 아르헨티나 등에 콩이 재배됨(20세기 후반)
- 아프가니스탄 전역에 콩이 재배됨(2013)

---

1) 이철호, 권태완, 콩의 이용역사, 콩, 한국콩박물관건립추진위원회편, 고려대학교출판부, 3-44(2005)

2) 김석동, 이영호, 콩의 품종과 육종, 콩, 한국콩박물관건립추진위원회편, 고려대학교출판부, 137-216(2005)

작물학계의 입장

식물학에서 작물의 발상지를 추정할 때 중요한 지표로 쓰이는 것이 야생종의 분포 유무다. 학계에서는 야생콩, 중간형, 재배콩이 한 곳에서 자라나는 곳을 식물의 원산지로 규정하고 있는데 이에 부합되는 곳이 바로 한반도와 만주일대다. 하지만 '한반도 원산지설'이 세계 학자들에게 주목받지 못한 이유 중 하나는 한국 콩에 대한 분석자료가 많지 않다는 점도 있다. 예를 들어 미국의 한 연구자가 중국이나 일본의 콩(유전자료)자료는 많이 이용하면서 한국의 (콩)자료는 누락하거나 소량만 이용한다면 결과는 아무 것도 없거나 미미한 수준일 것이다.

최근 서울대 이영호 박사는 한반도 남부에서 속속 출토되고 있는 청동기시대의 탄화 콩과 콩의 유전적 다양성을 종합 정리하여 발표한 바 있다. 그는 "지구의 최대 빙하 발달기 이후(기원전 11000년) 중국대륙과 한반도가 분리된 후 한반도에서 독자적으로 한반도인의 선택과 순화에 의해 재배두류가 발달되었다"고 하였다. 서울대 이석하 교수도 "앞으로 두류의 유물이 추가 발굴되고, 콩의 진화를 규명할 수 있는 DNA분석에 의한 분자시계(molecular clock)를 규명하게 되면 남만주와 한반도가 콩의 식물학적 기원지로 널리 인정받을 수 있을 것"이라고 보았다.[1]

고고학계-출토유물은 말한다

안승모 원광대 교수는 "신석기시대부터 야생으로 추정되는 두류가 이용되고 있지만 확실한 재배종은 기원전 1천 년 대의 유적에서부터 집중적으로 출토된다. 고고학적으로 보아 콩은 만주지방과

---

1) Kim, M. Y. et al., Whole-genome sequencing and intensive analysis of the undomesticated soybean (Glycine soja sieb and Zucc.) genome, PNAS, 107(51), 22032-22037(2010)

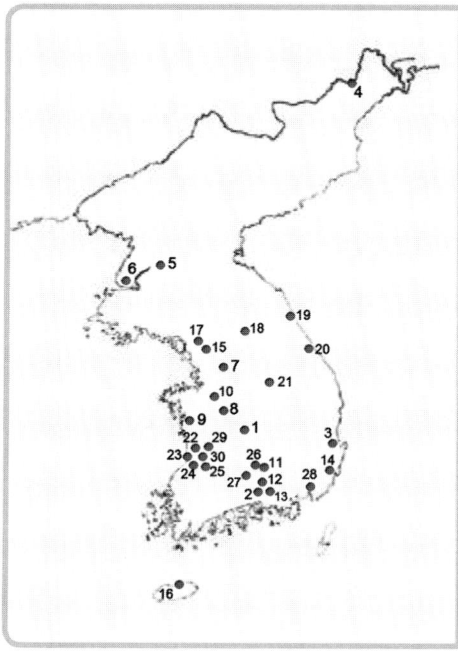

유적	시대
1. 옥천 대천리 유적 2. 진주 상촌리 유적	신석기 시대
3. 포항 원동 유적 4. 회령 오동 유적 5. 평양 남경 유적 6. 황해도 석탄리 유적 7. 양평 양근리 유적 8. 청원 궁평리 유적 9. 보령 평라리 유적 10. 천안 백석동 유적 11. 합천 봉계리 유적 12. 진양 대평 유적 1 13. 진양 대평 유적 2 14. 울산 다운동 유적	청동기 시대
15. 고양 가와지 유적 16. 제주 삼양동 유적	초기철기 시대
17. 파주 주월리 유적 18. 횡성 둔내 유적 19. 양양 가평리 유적 20. 명주 안인리 유적 21. 단양 수양개 유적 22. 군산 관원리 유적	원삼국 시대
23. 부안 반곡리 유적 24. 부안 토산리 유적 25. 김제 심포리 유적 26. 김해 부원동 유적 27. 산청 산남리 유적 28. 합천 저포C지구유적 29. 논산 원북리 유적 30. 익산 왕궁리 유적	삼국 시대

한반도 출토 콩 관련 유적

한반도에서 기원한 것으로 추정된다. 늦어도 3천 년 전 경에는 한반도와 길림지역의 청동기유적에서 콩이 출토되고 있는 반면 중국은 주로 전한시대(BC 202~AD 220)부터, 일본은 기원전 4세기경 이후의 야요이 시대부터 발견된다."고 하였다.[1)]

한반도 두류 관련 출토 유물

한반도 두류관련 출토유물을 살펴보면 신석기시대, 청동기시대, 초기철기시대, 원삼국시대, 삼국시대 등에 걸쳐 많은 콩 유적이 수십 곳에서 발견되고 있다.[2)]

---

1) 조현종, 선사 고대 유적 중의 콩, 콩, 한국콩박물관건립추진위원회편, 고려대학교출판부, 45-80(2005)

2) 안승모, 두류재배 기원에 대한 고고학적 고찰, 한국콩연구회지, 19(2):24-33(2002)

신석기시대(BC 1500 이전)

콩과 관련된 신석기시대의 자료는 옥천 대천리 유적과 진주 상촌리 유적이 있다. 그러나 대천리 유적에서 탄화미와 보리, 밀, 조 등과 함께 보고된 콩자료는 정식보고서에는 언급되지 않고 있으나 경남 진주의 상촌리 유적에서 밀, 보리, 조, 기장 등의 곡물자료와 함께 콩과식물이 보고됨으로써 이 유적이 우리나라 콩 자료의 최고 유적이 되었다.[1)]

청동기시대(BC 1500~BC 300)

청동기시대에는 벼농사가 본격화되고 오곡이 완성되는 등 우리 식문화의 전통이 확립된 것으로 알려지고 있다. 청동기시대의 콩류는 함북 회령 오동, 평양 남경, 황해 석탄리, 경기 양평 양근리, 충북 청원 궁평리, 충남 보령 평라리, 충남 천안 백석동, 경남 합천 봉계리, 경남 진양 대평, 울산 다운동 등 12개 유적에서 확인되고 있다.

주거지 10호 전경

출토된 콩류와 덩어리

포항 원동 청동기시대 주거지 10호 전경[2)]

---

1) 이영호, 박태식, 출토유물과 유전적 다양성으로 본 한반도의 두류 재배 기원, 농업사 연구 제5권 1호, 1-31(2006)

2) 박태식/이영호, 포항 원동 유적(2003:282)

초기철기시대(BC 300~0)

철제 농기구의 보급은 농업생산력을 급속히 증가시키는 결과를 가져왔다. 이 시기의 콩 자료는 김포 고양 가와지 유적과 제주 삼양동 유적이 있다. 콩은 가와지 유적에서는 벼, 박, 복사나무씨 등과 삼양동 유적에서는 탄화미, 밀, 보리 등과 같이 발견되었다.

원삼국시대(AD 0~300)

파주 주월리, 횡성 둔내, 양양 가평리, 명주 안인리, 단양 수양개, 군산 관원리 유적 등 6개 유적에서 확인되고 있다. 종류는 콩과식물의 콩, 팥, 녹두가 확인되고 있으며 출토된 유물은 모두 주거유적에 속한다. 동부를 제외하고 각 유적에서 출토되고 있는 콩과 팥, 녹두는 청동기시대 이래 콩과식물의 세트화가 이루어지고 계속적으로 재배되었음을 의미한다.

삼국시대(AD 300~668)

부안 반곡리와 토산리, 김제 심포리, 논산 원북리, 익산 왕궁리, 김해 부원동, 산청 소남리, 합천 저포리 지구 등 8개 유적에서 검출되었다. 출토된 곡류 가운데 원북리 유적의 출토 새팥의 경우는 크기가 비교적 작은 야생콩류로 이것은 콩과의 작물이 재배되고 있는 동안에도 여전히 야생 두류가 식용되고 있음을 알 수 있는 자료이다.

3. 시대별, 콩의 모습

고조선이 실제 어떻게 존재했는지, 또 얼마큼 부강한 나라였는지 별로 알려진 것이 없다. 다만 우리는 청동기시대의 유물들 즉 고인돌과 비파형동검, 빗살무늬토기 등을 통해 그들의 문화수준을 짐작

할 뿐이다. 고조선은 이후 부여와 고구려로 이어졌다. 우리 민족의 간판이 고조선, 고구려, 발해, 통일신라, 고려, 조선 등 그 이름이 무엇이었든 간에 콩은 수천 년간 우리 민족의 운명과 함께한 '민족의 음식(National Food)'이었다.

고조선 시대

갑골문의 두(豆)

콩을 의미하는 한자로 한자문화권에서 두루 쓰이는 글자는 '두(豆)'다. '두(豆)'가 제일 먼저 선보이는 곳은 은나라 유적지에서 발견된 갑골문에서다. 은나라를 건국한 주역은 동이족이라고 알려져 있다. 은나라 시대(BC 1600~BC 1046), 중국의 동북부에는 고조선(BC 2333~108)이 존재하였다. 갑골에 처음 등장하는 '두'는 콩이 아니라 '제기(祭器)'를 의미한다. 제사지낼 때 쓰는 굽다리그릇을 의미하는 '두'가 언제부터 콩을 의미하게 된 것일까. 3~4천 년 전 고대인들은 수렵, 채집의 시대를 마감하고 정착생활을 하기 시작하

갑골문

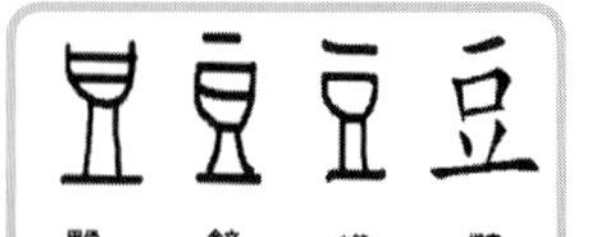

중국 은나라에서 점을 치기 위해 거북이 등껍질이나 소의 어깨뼈를 이용해 새겼던 문자. 1899년 베이징의 관리였던 왕의영은 어느 날, 학질을 다스리기 위해 사용하던 약재에 거북 뼈 조각이 섞여 들어가 있다는 것을 발견했다. 약 3천 년 전 은나라 유적지에서 출토된 은대 후기경의 갑골에 새겨진 글을 한자의 원형으로 보고 있다.

였다. 땅에서 얻은 수확물은 당연히 신에게 드리는 감사의 제물이 되었을 것이다. 이는 아메리카 신대륙에서 첫 추수를 앞두고 감사의 예배를 드렸던 '추수감사절'의 유래와도 비슷하지 않았을까. 우리가 의식을 하든 못하든, 또 알든 모르든 콩은 제사, 제물, 신, 조상, 감사 등의 의미와 중첩되어 있다. 콩의 중요성의 무게는 결코 가볍지 않다.

융숙(戎菽)

동북아에서 문헌상으로 보면, 콩은 「시경詩經」에 숙(菽)이라는 이름으로 처음 등장한다. 여기에 나오는 '두'는 제기, '숙'은 콩의 의미다. 그런데 한나라 무렵, 숙의 꼬투리가 나무로 만든 제기인 두(豆)와 비슷하다 하여 숙(菽)은 두(豆)가 되어버렸다. 콩을 '두'라고 한 것은 한나라 때, 즉 서기 전후의 일이다. 콩이란 의미로 역사책에 제일 먼저 기록이 된 것은 기원전 7세기 관중이 쓴 「관자」에서다. 제나라 군사들은 동이(산융)의 콩(戎菽)을 가져갔다. 콩을 갖고 갔다는 것은 들판에 아무렇게나 자라는 야생콩은 아니었을 것이다. 기원전 천 년 전의 콩(융숙)을 거론한 역사서 등을 볼 때, 제나라가 동이(산융)을 쳐들어왔을 때는 이미 야생콩을 순화시킨 재배콩이 자라고 있었던 것은 아니었을까. 어쩌면 콩 때문에 전쟁이 벌어졌는지도 모를 일이다.

고구려시대

고구려 선장양

고구려의 발달된 발효기술은 인근 나라에 '고려취'라 하여 이미 알려졌었다. 3세기말 장화(232~300)는 박물지에서 "외국시유법

(外國豉有法), 즉 외국에 메주가 있다"라고 하였다. '고구려선장양(高句麗善藏釀)'이란 말은 「삼국지위지동이전」(3세기경)에 나오는 말로, "고구려 사람들은 발효식품을 잘한다"라는 뜻이다. 여기서 유의 깊게 보아야 하는 것은 중국 최고의 농업기술서라고 하는 「제민요술(532)」에서의 기록이다. 대두의 종류를 4가지로 언급하고 있는데 '고구려의 노란콩(高麗黃豆)'와 '고구려의 검은콩(高麗黑豆)', '비두(飛豆 : 제비콩)', '완두(豌豆)'라 하였다. 비두와 완두는 대두가 아니다. 6세기 무렵 '대두는 바로 고구려 유래의 콩밖에 없었다.

덕흥리 고분의 간장

"무덤을 만드는데 1만 명의 공력이 들었고, 날마다 소와 양을 잡아서 술과 고기, 쌀은 다 먹지 못할 정도이다. 아침에 먹을 간장(豆醬)을 한 창고 분이나 두었다. 무덤 찾는 이가 끊어지지 않기를..." 덕흥리 고분은 남포 특급시 강서구역 덕흥동에 있는 고구려시대의 고분으로 북한의 국보 문화유물 제156호로 지정되어 있다. 이 고분은 고구려의 대신급 인물이었던 '진(鎭)'의 무덤으로, 408년(광개토대왕 17)에 축조되었다. 벽과 천장에는 여러 가지 내용의 인물풍속도와 이를 설명하는 600여 자의 글자가 있고 409년 2월 2일을 마지막으로 무덤의 문은 닫히게 된다. '진'의 국적을 두고 중국인이다 고구려인이다 논란이 되고 있으나, 확실한 것은 5세기 초의 고구려 문화와 풍습을 살펴볼 수 있는 귀중한 유적이라는 것이다. 간장을 한 창고분이나 남겨둔 채 무덤의 주인은 역사 속으로 사라져갔다. 이때 고구려는 광개토대왕이 집권하던, 최고의 전성기를 누리던 시절이었다.

통일신라시대

신문왕 때의 폐백음식

삼국사기 신라본기편에 보면 신문왕 3년(683)에 왕이 김흠운의 딸을 부인을 맞이할 때, 폐백품목으로 쌀(米), 주(酒), 유(油), 밀(蜜), 장(醬), 시(豉), 포(脯) 등 총 135수레를 보냈다는 내용이 있다(弊帛十五擧 米酒油密醬豉脯 一百三十五擧) 이로써 그 당시에도 (왕이) 폐백품목으로 보낼 만큼 시와 장류는 중요하게 취급되었음을 알 수 있다. 여기에 나오는 '시'는 지금으로 말하면 메주다. "메주의 역사는 곧 장(醬)의 역사이다"라고 할 만큼, 메주는 우리나라 전통장의 발효원으로 항상 중요하게 취급됐다.

삼국유사 : 흥륜사의 장을 발라라

처녀로 변신한 호랑이가 김현(金現)과 부부 인연을 맺은 뒤 그를 위해 죽음을 택했다는 감동적인 설화가 삼국유사(BC 57~935) 김현감호(金現感虎)편에 전해진다. 내용은 김현이라는 사람이 호랑이를 감동시킨 이야기로 "호랑이에 물린 사람들은 흥륜사의 장을 발라라..."라는 말이 나온다. 얼마 전까지도 우리는 민간 처방으로 벌에 쏘이는 등 유사시엔 된장을 바르곤 했다. 이를 볼 때 훨씬 오래전부터 우리 민족은 민간약으로 구하기 쉬운 된장을 이용해왔음을 알 수 있다. 흥륜사는 법흥왕(528년)이 불교를 공인한 뒤 세워진 신라 최초의 절로 진흥왕(544년)때에 완성되었다. 진흥왕은 이 절을 '대왕흥륜사'라 하였고, 나중에 이 절의 주지가 되었다. 조선시대엔 절메주로 궁궐의 진장을 담았다하고, 또 왕릉에는 두부를 만드는 조포사가 딸려 있었던 것을 볼 때 사찰과 콩, 또 왕실과 콩은 아주 오래전부터 불가분의 관계를 맺고 있었던 것 같다.

발해시대

책성의 메주

발해의 유명한 생산품으로는 '책성의 시(豉)' 가 있었다. '시'의 해석을 두고 의견이 분분한 데, 된장과 간장을 가르지 않은 초기의 장이거나 장을 담그기 전의 누룩 상태 혹은 메주로 보는 의견이 가장 많다. 책성(柵城)은 오늘의 길림성 훈춘현 팔련성 일대인데, 예나 지금이나 콩의 명산지에 해당된다. 발해 때의 농작물의 품종은 기후와 토양 지세에 따라 달랐는데 기후가 따뜻한 지역에서는 조, 콩, 벼를 기온이 낮은 지역에서는 조, 콩 등이 재배되었을 것으로 보고 있다. 「신당서(新唐書618~907)」에서는 '귀하게 여기는 것은 책성의 메주(柵城之豉)'라고 하였다.

'시(豉)'는 'soy'인가

「예기(禮記 BC 450~AD 100)」나 「초사(楚辭 BC 300)」에도 '시란 무엇인가'라고 적고 있다. 진대(秦代 BC221~209)이전의 중국 문헌에는 시에 관한 언급이 없으나 서한 시대에는 시가 주요식품이었다. 「사기(史記 BC 90)」에서는 얼국과 염시가 1천독이라고 하였다. 진대(晋代) 장화의 「박물지(博物志 AD 190)」에서는 "외국에 시가 있다"라고 했으며 이시진(1518~1593)의 「본초강목」에서도 시는 외국 원산이라고 하였다. 송대(宋代 960~1126)의 「학제점필(學齊佔畢)」에서도 구경(九經) 속에 시豉란 자가 없고 방언에 시豉가 있을 뿐이다"라고 하였다. 고구려를 계승한 발해가 '시(豉)'로 유명했고 통일신라 초기에도 '시(豉)'는 존재했다. 하지만 조선시대에서는 메주가 널리 통용되고 청국장 혹은 전국장이 나오면서 '시豉'는 사라지게 되었다. '시'는 현재의 중국어로 읽으면 '쓰' 혹은 '쯔'가 되지만 중국 고대어로 읽으면 '싀'가 된다는 것이다. 우리나라 중세국어사전에서도 '시'는 '싀'로 되어 있다. 영어로 콩은 'soybean'이다. 일본의 간장인 '쇼유(醬油)를 만드는 콩(bean)이라는 말이다. '시', '싀', '쇼', '소이'의 변화가 예사롭지 않다.

고려시대

구황식품(굶주림 구제식품)

고려사(1058) 식화지(食貨志)의 기록을 보면, 1018년(현종 9년)에 거란의 침입으로 굶주림과 추위에 떠는 백성들에게 쌀, 소금, 장이 지급되었다는 것과, 1052(문종 6년)에 개경의 굶주린 백성 3만 여 명에게 쌀, 조, 된장을 내렸다는 기록이 있다. 굶주린 백성을 구제하기 위한 구황식품은 생명 유지를 위해 필요한 기본식품이다. 이 구황식품에 쌀, 조와 같은 곡물과 함께 장과 된장이 들어 있었다는 것은 고려시대에 이미 장류가 필수식품으로 정착되었음을 말해주고 있다.

조선시대

장의 전성기

조선시대 최고의 조리서인 『규합총서1809』에는 장 제조법뿐 아니라 장 담그는데 따른 택일법, 금기사항, 보관 관리법 등이 기록되어 있다. 간장은 부종이 일어나지 않게 하고 기력을 유지하는데 효과가 있다 하여 「구황촬요救荒撮要」「구황절요救荒切要」 등 구황식품서에도 콩잎을 이용한 흉년기의 장제조법을 자세하게 기록해 두었다. 「요록要綠1680」, 「주방문酒方文」, 「시의전서是議全書」 등 조선시대 조리서에도 대부분 간장 제조법이 적혀 있다. 「산림경제」에는 25조목이, 「증보산림경제 1766」에는 45조목의 장 제조법이 적혀 있다. "장은 모든 맛의 으뜸"이라는 뜻으로 "장(醬)은 장(將)이다" 또 "가장은 모름지기 장 담그기에 뜻을 두어 오래 묵혀 좋은 장을 얻어야 할 것이다"라는 내용이 있다.

「증보산림경제」에서 말하는 장제조법

별장법, 침장법, 장담글 때 여러 재료를 섞는 법, 장 담글 때 피해야 할 점, 장독을 잘 두는 법, 맑은 장을 뜨는 법, 장이 맛을 잃었을 때 되살리는 법, 생선과 고기를 넣어 장 담그는 법, 생황장법, 숙황장법, 메밀로 장 담그는 법, 보리로 장 담그는 법, 느릅나무열매로 장 담그는 법, 팥으로 장 담그는 법, 청태콩으로 장 담그는 법, 장을 빨리 만드는 법, 맑은 장을 빨리 만드는 법, 고추장 만드는 법, 고추장을 빨리 만드는 법, 즙장의 메주를 만드는 법, 여름철에 즙장 담그는 법, 전시장을 만드는 법(속칭 戰國醬), 청태전시장법, 수시장법, 달걀장법, 장을 볶는 법, 장을 달이는 법, 장 떡 만드는 법, 담수장법, 천리장법 등이다.

근대

방곡령

'식량의 무기화'는 식량에 대한 위기를 말할 때, 가장 많이 나오는 말이다. 근대 개화기 무렵 우리나라에서 일어난 방곡령 사건을 보면 식량과 전쟁이 어떤 식으로 엮이게 되는지 알 수 있다. 1889년 함경도 관찰사 조병식은 일본인들에게는 아예 쌀과 콩을 팔지 말도록 방곡령을 발포하였다. 일본은 방곡령의 철회와 손해배상을 요구하였고, 식량부족에 대한 농민들의 불만은 동학봉기로 이어졌다. 대원군은 중국에 도움을 청하고, 명성황후는 일본군에 도움을 청하면서 우리 땅에서 청일전쟁이 일어났다. 결과는 일본의 승리로 끝나면서, 쌀과 콩 등의 곡식수탈로 이어졌다. 1917년부터 1928년까지 12년 동안 일본으로 유출된 곡식은 우리나라 전체 생산량의 절반에 해당하였다. 일본은 1931년 만주사변을 일으키고 이후 3년간 만주를 지배했다. 1933년 진주만을 폭격하고 제2차 세계대전을 일으킨 일본이 어디서 군량미를 확보했있는지 짐작케 하는 대목이다.

세계대전과 콩

콩이 20세기 들어와 전 세계로 확산되고 세계의 주요 곡물이 된 데에는 제1차 세계대전과 제2차 세계대전의 영향이 컸다. 1차 대전을 겪으면서 서방 세계는 극심한 식량난을 경험하게 된다. 영국은 1918년 식량배급제를 실시하고 콩가루와 감자가루를 섞어 빵을 만들도록 했다. 1918년 미국에서도 "밀, 고기, 지방을 절약하기 위해 콩가루를 쓰자"라는 운동을 벌였다. 오늘날 미국이 세계 제1의 콩 생산대국이 된 것은 "전쟁에 이기기 위해서는 콩을 심어야 한다"며 콩 생산지역을 빠르게 확대해나간 결과라고 볼 수 있다. 현재 우리나라의 콩의 자급률은 10%가 채 안 된다. 유사시엔 쌀은 보리나 밀로 대체하고 고기는 콩으로 대체할 수 있다. 하지만 콩을 대체할 수 있는 작물이나 식품은 없으니 콩은 중요한 무기인 셈이다.

4. 인물로 본 콩 이야기

콩과 관련된 동서양의 유명 인물 33인을 선정하였다. 어떤 사람은 콩에 관해 아무런 공헌이 없지만 콩의 역사에 있어 중요한 일과 겹쳐 있거나, 그 인물이 살았던 시대배경과 맞물려 있기 때문에 선정하였다. 그러므로 여기에 나온 인물은 세계적으로 콩의 전파와 확산에 기여했다하겠다. 시대가 분명하게 언급된 경우도 있지만, 정확히 언제 일어난 일인지 알 수 없는 경우에는 그 인물이 태어난 연대순으로 나열하였다.

제나라 환공(재위 BC 685~643) : 콩, 중국에 보급

콩이 역사책에 처음 등장한 것은 제나라 환공 때였다. 제나라의 환공은 춘추오패의 첫 번째 패자(覇者)였다. 그의 곁에는 중국을 통틀어 최고의 재상으로 꼽히는 관중이 있어 그가 최고의 자리에 오를 수 있도록 도왔다. 관중과 그 제자들이 쓴 「관자管子」에 바로 '제환공이 산융을 치고 고죽국까지 갔다가 융숙을 가져온' 이야기가 있다.

앞장에 설명한 바와 같이 융숙(戎菽)은 산융의 콩이다. 제 환공은 연나라의 요청을 받고 연나라를 괴롭히는 북방의 산융, 고죽국을 치고 돌아오는 길에 콩을 가지고 와 중국에 퍼뜨린 것으로 보인다.

공자(BC 551~BC 479) : 「시경」에서 콩을 노래하다

3, 4천 년 전 발굴된 탄화 콩의 유적을 감안해보면 콩의 재배는 제나라 훨씬 이전부터 이루어졌을지도 모른다. 콩을 의미하는 '숙'과 '두'란 글자는 2500년 전, 공자가 편찬한 「시경」에 나온다. 이때 콩을 의미하던 글자는 '숙(菽)'이었고, '두'는 '제기'의 의미였다. 주나라 무왕을 노래한 '생민(生民)'이라는 시에는 '콩을 심으셨는데 콩은 너풀너풀 자라났고,(蓺之荏菽 荏菽旆旆)' '제기에 제물을 담는데 접시도 있고 대접도 있네(卬盛于豆 于豆于登)' 라고 하였다. 오래 세월이 지난 후, 제기(豆)는 그 위에 자주 올려 지던 제물(豆)과 동일시되어 '콩'이 되었다.

유안(BC 178~122) : 두부의 시조

유안은 한나라(BC 202~AD 220)를 세운 유방의 손자로 회남왕에 봉해진 인물이다. 우리나라에서 두부가 처음 선보이는 곳은 당말송초(唐末宋初) 때의 문헌인 「청이록淸異錄」에서다. 북방 유목민족과 교류가 있었던 당나라(618~907) 후기에 유락(乳酪)문화의 영향을 받아 두부가 들어왔을 것으로 보고 있다. 그런데 명나라 때의 이세진이 「본초강목 1596」이란 책에서 '두부를 발명한 사람은 유안'이라고 기록하면서부터 유안은 '두부의 시조'로 이름을 떨치게 되었다. 혹자는 두부 발명을 했다는 시기와 문헌상 기록의 차이가 무려 1500년이나 되는 것을 두고 의구심을 나타내기도 한다.

조식(192~232) : 자두연기

‘자두연기(煮豆燃萁)’는 ‘콩을 삶는 데 콩깍지를 태운다’는 뜻으로, 형제간의 다툼을 비유하는 말이다. 「삼국지」 최후의 승리자인 조조에게는 조비, 조식 두 아들이 있었다. 막내인 조식은 뛰어난 문재(文才)로 조조의 사랑을 독차지하였다. 조비에게는 항상 조식을 경계하는 마음이 있었다. 제위는 조비에게 돌아갔다. 어느 날 조비는 조식을 불러 일곱 발걸음을 내딛는 동안 시를 한 수 짓지 못하면 안된다고 으름장을 놓았다. 이때 조식의 나이 열 살이었다. 어린 나이에 아래의 시를 지었다고 하니 사물(콩)을 보는 관찰력이 대단하다하겠다.

> 콩을 삶는데 콩깍지로 불을 때니[煮豆燃豆],
> 콩이 솥 안에서 우는구나[豆在釜中泣].
> 본래 같은 뿌리에서 나왔거늘[本是同根生],
> 왜 서로 들볶아야만 하는지[相煎何太急]

배현경(874~936) : 콩나물 일화

우리 민족이 콩나물을 먹기 시작한 지는 꽤 오래되었을 것으로 보이나, 콩나물이 공적으로 언급되기 시작한 때는 고려시대 초기이다. 935년 고려의 태조가 나라를 세울 때 태광대사 배현경이 식량부족으로 허덕이던 군사들에게 콩을 냇물에 담가 콩나물을 만들어 배불리 먹게 했다는 일화가 전해진다. 문헌상으로는 고려 고종 때(1214~1260)의 「향약구급방」에 대두황권으로 기록된 것이 콩나물에 대한 첫 기록이다. 대두황권은 싹 튼 콩나물을 햇볕에 말려서 약으로 사용하는 것이다.

이색(1328~1396) : 「목은집」에 두부

두부가 문헌에 처음 등장하는 때는 고려 말이다. 당시 가장 교류가 많았던 원나라로부터 전래되었을 것으로 보고 있다. 두부가 어떤 경로로 우리나라에 전해졌는지 정확히 알 수는 없지만 이색이 쓴 「목은집」에 처음 소개 된다. 이색은 일찍이 원나라에 유학하고, 거기서 관직을 역임하고 돌아온 해외파다. 그는 두부를 아주 좋아했었는지, 「목은집」에 여러 편의 두부 관련 시를 남겨두었다. "두부 반찬에 토란을 곁들이었고" "기름에 두부를 튀겨 잘게 썰어서 국을 끓이고"란 내용도 있다. '두부가 마치도 금방 썰어낸 비계 같군' 또는 "성긴 이로 먹기에는 두부가 그저 그만"이란 내용도 있다.

이성계(1335~1408) : 고추장 전설

고추가 유입된 시기는 16세기 임진왜란 전후로 보고 있다. 하지만 조선의 태조 이성계와 관련된 순창고추장 전설이 전해지고 있으니 흥미롭다. 과연 이성계가 먹은 것이 지금과 같은 고추장이었을까. 이성계는 왕위에 오른 후, 스승인 무학대사가 기거하고 있던 순창군 소재 만일사를 찾았다가 어느 농가에 들러 고추장을 곁들여 점심을 먹었다. 환궁한 이후에도 그 맛을 잊지 못하여 진상하도록 한데서 고추장 전설이 탄생하였다. 「소문사설1740」에 전복, 큰새우, 홍합 등을 넣은 '순창고초장조법'이 전해지고 있으니 순창고추장에는 뭔가 특별한 것이 있기는 있는 모양이다.

세종(1397~1450) : 두부 만드는 여자, 명나라에 파견

두부가 원나라로부터 전해졌다는 시기는 12세기 무렵이다. 하지만 아이러니컬하게도 15세기 세종 때에는 우리나라의 '두부 만드는 여자들이' 중국 황실에서 인기가 높았다. 세종 10년(1428), 사신으

로 간 관리는 명나라 황제에게 두부 등을 올렸다가 벼슬을 받아왔다. 또 세종 16년(1434)에는 사신이 중국 황제의 칙서를 가지고 왔는데 "먼젓번에 보내 온 반찬과 음식을 만드는 부녀자들이 모두 음식을 조화(調和)하는 것이 정하고...두부를 만드는 것이 더욱 정묘하다... 왕이 다시 여자 10여 인을 뽑아 보내 달라"는 내용이었다. 당시, 조선에는 두부를 두포(豆泡)라고 했고, 포장(泡匠)이라는 직책이 있었다.

세종, 콩으로 정치하다

「조선왕조실록」을 보면 "세종은 콩으로 정치했다"고 할 정도로 콩을 자주 사용했다. 세종은 32년 재위기간 동안 총 566건이나 콩에 대해 언급했다. 41년 재위한 선조는 178건, 52년 재위한 영조는 73건이었다. 세종 때는 쌀이 거론된 횟수도 다른 왕보다 훨씬 많았다. 세종 893건, 선조 511건, 영조 591건이었다. 더 놀라운 것은 쌀에 대한 콩의 비율인데 세종은 63%, 선조 35%, 영조 12%였다. 그만큼 세종은 식량의 소중함을 알고, 또 특히 콩을 사랑했던 왕으로 여겨진다. 세종 때 언급된 콩의 용도를 살펴보면 세쌍둥이 축하선물, 흉년구제용, 콩씨(종자용), 말먹이, 부조, 교역, 녹봉, 조세, 지세, 코끼리먹이, 종친하사품, 환상곡, 난민구제, 대출용, 회유책 등 그 용도가 다양하였다.

문종(1450~1452) : 간수의 문제

문종 1년(1451), 조선 왕실에서는 두부제조와 관련, 회의가 열렸다. 염전을 소로 갈기 때문에 정갈하지 못하다는 것이 문제였다. 특히 제향(祭享) 및 공상(供上)하는 두부는 산수(酸水)를 써야 한다"고 말했다. 문종은 "두부를 만드는 데 무슨 물을 쓰느냐? 어떤 이는 소금의 융액(融液)을 쓴다 하고 어떤 이는 바닷물을 쓴다 하니, 누구의 말이 옳은지 알지 못하겠다."고 하였다. 당시 조정에서는 두부를 만들 때, 비위생적인 간수를 계속 써야 하는지 궁금해

했다. 두부 단백질은 칼슘과 마그네슘 등에 의해 응고가 된다. 식초에 약간의 소금을 더한 산수로도 두부는 응고된다. 600년 전 논쟁에서도 배울 것이 있다.

강희맹(1424~1483) : 서리태, 서목태의 기록

조선의 공식적인 문건에서는 콩을 이르는 말로 '숙' 혹은 '두'를 쓰고 있었다. 하지만, 민간에서는 서리태, 서목태 등 '태'도 통용되었다. 조선 성종 때의 문신이었던 강희맹은 은퇴한 후 경기도 금양현(지금의 과천)에 살면서 「금양잡록1492」이란 책을 썼다. 강희맹은 직접 농사도 짓고 시골의 농로들과 어울리게 되면서 민간의 말을 그대로 적었던 것으로 보인다. 이 문헌에서 처음으로 서리태, 서목태, 황태, 흑태가 선보였다. 그 이후, 여러 농서에서는 콩을 의미하는 글자로 '두'와 '태'를 병용하였다. 우리는 요즘도 지역에 따라 용도에 따라 또 연령에 따라 콩, 두, 태를 두루 사용한다. 백태, 흰콩, 노란콩, 장콩, 메주콩, 검은콩, 서목태, 서리태 등은 모두 '큰콩(대두)'을 의미한다.

허엽(1517~1580) : 초당두부

16세기 중엽 초당 허엽이 강릉 부사로 있을 때였다. 당시 관청 앞마당에 샘물이 있었는데 물맛이 좋았다. 그는 우물물로 두부를 만들고 바닷물로 간을 맞췄다. 두부는 맛이 좋기로 소문이 났다. 허엽은 자신의 호를 붙여 '초당' 이름 짓고 이 두부를 팔아 큰돈을 벌었다고 한다. 샘물이 있던 자리는 현재 강릉시 초당동이 되었고, 지금도 그 곳에는 허엽을 기리는 비석이 있다. 허엽은 허균과 허난설헌의 아버지다.

도꾸가와 이에야스(1543~1616) : 하초미소

집집마다 장을 담그던 우리나라와는 달리 일본의 경우는 마을 단위로 혹은 막부 중심으로 장을 담갔다고 한다. 보통 일본된장을 만들 때는 쌀이나 보리를 절반은 섞는데, 현재 나고야의 명물로 알려진 하초미소는 특이하게도 콩만을 가지고 2년 이상 숙성시켜 만든다. 일본에서도 하초미소는 맛이 깊고 특유의 향이 있어 최고의 상품으로 인정받고 있다. '하초미소'는 오카자키성과 '8정 = 하초'이 떨어져있다는 데서 유래되었다. 하초미소공장은 1337년부터 존재했다고 한다. 그렇다면 1542년, 오카자키성에서 태어난 도꾸가와 이에야스도 하초미소를 먹었으리라.

이순신(1545~1598) : 「난중일기」 연포탕

이순신 장군이 쓴 「난중일기」에는 조선시대 연두부국 요리인 연포탕에 대한 기록이 있다. 요즘 연포탕이라면 맑은 국물에 낙지를 살짝 데친 요리를 말하는데, 원래는 '맑은 두부국'을 의미했다. 명나라의 원군들이 조선에 왔을 때 두부가 중요한 부식으로 제공되었다. 예를 들면 중군·천총·파총에게는 '고기 + 두부 + 소채 + 절인 생선 + 밥 + 술', 각 아문의 차인에게는 '고기 + 두부 + 소채 + 밥', 군병에게는 '두부 + 절인 새우 + 밥'을 지급하였다고 한다. 두부는 모두에게 제공된 공통 메뉴였다.

허균(1569~1618) : 두부를 잘 만드는 사람들

1611년 허균이 쓴 「도문대작屠門大嚼」은 '조선 최초의 음식품평책'이라고 할 수 있다. 이 책은 허균이 전라북도 함열로 귀양 가 있던 시기에 저술한 것이다. 유배지에서 거친 음식만을 먹게 되자, 예전에 먹었던 맛있는 음식을 차례대로 떠올리며 서술한 책이다.

'두부'라는 항목 중에는 "장의문 밖에 사는 사람이 두부를 잘 만드는데 연하고 매끄러운 것이 무엇이라 말할 수 없다"고 했다. 도문(屠門)이란 소나 돼지를 잡는 푸줏간의 문이고, '대작(大嚼)'은 크게 씹는다는 뜻이다. 즉, 도문대작은 "현재 먹을 수 없는 고기를 생각하며 푸줏간 문을 향해 입맛을 다신다" 라는 제목이라니, 과연 홍길동의 저자다운 발상이 아닐 수 없다.

루이14세(1638~1715) : 일본간장

프랑스가 요리로 이름나게 된 것은 16세기 이탈리아의 카트린 메디치와 프랑스의 앙리 2세가 결혼하면서 부터다. 즉 이탈리아요리가 프랑스에 접목되면서 프랑스는 '요리의 전성기'를 맞게 된다. 특히 루이 14세와 루이 15세는 미식가로 이름이 높았는데 이들을 만족시키기 위해 갖은 요리법이 개발되었다. 일본의 간장이 네덜란드 동인도회사를 통해 전해진 것도 이 무렵이었다. 간장은 주로 샐러드용 드레싱을 만드는데 이용되었다. 이 간장소스는 간장, 레몬주스, 다진 마늘, 고춧가루, 올리브오일, 설탕, 식초 등을 만들었다. 간장은 루이 14세가 집권하던 시절, 프랑스 귀족들 사이에서 최고의 향신료 중 하나로 이용되었다. 일본간장은 동양의 콩식품 중에서 가장 먼저 서양에 알려지게 되었다.

이익(1681~1763) : 삼두회

초기 실학파의 한 사람이었던 성호 이익은 「성호사설」에서 "사람을 살리는 데는 콩의 힘이 가장 크다"라고 간파한 바 있다. 이익은 72세 때 콩을 직접 재배하면서 '삼두회(三豆會)'를 만들었다. '삼두회'는 콩으로 만든 3가지 즉 콩죽, 콩나물, 된장을 먹는 모임이었다. 친인척이 한데 모여 좋은 글도 돌려 읽고 또 생활규범을 공부하는

자리였다. 성호는 경기 일대에 자주 흉년이 들어 백성들의 삶이 피폐하게 되자, 일찍이 구황 정책이나 구황 작물에 남다른 관심을 보였다. 그 중에서도 콩을 매우 중요한 대안으로 생각하고, 그 활용법에도 적극적이었다. 성호는 '콩을 먹는 보통 사람들의 나라 걱정'이란 뜻으로 「곽우록藿憂錄」을 쓰기도 했다. 성호 이익이야말로 조선을 통틀어 콩의 가치를 제대로 인식한 최고의 학자라 하겠다.

영조(1694~1776) : 고추장을 사랑한 임금

조선의 임금 중 가장 긴 재위기간(52년)을 보냈던 영조는 유독 식성이 까다로웠다. 승정원 일기에 따르면 고추장에 대해 언급한 기록(1749년 7월 24일)이 있다. "옛날에 임금에게 수라를 올릴 때 반드시 짜고 매운 것을 올리는 것을 봤다. 그런데 지금 나도 천초(川椒·산초) 등과 같은 매운 것과 고초장을 좋아하게 됐다." 영조는 내의원에서 만든 고추장이 사대부 집안에서 만든 것만 못하다고 평가하였다. 또 "송이, 생복, 아치(어린 꿩), 고초장(苦草醬) 이 네 가지 맛이 있으면 밥을 잘 먹으니, 이로써 보면 입맛이 영구히 늙은 것은 아니다"라고도 하였다.

벤저민 프랭클린(1706~1790) : 미국에 두부를 전하다

미국에 처음 콩을 들여온 사람은 중국을 다녀온 사무엘 보웬이란 선원이었다. 하지만 미국 땅이 넓은데다 통신이 발달하지 않았을 때는 어느 주에서 무슨 일이 일어났는지 전혀 알 수 없었을 것이다. 미국 건국의 아버지 중 한사람인 벤저민 프랭클린은 프랑스의 대사로 있을 때, 콩을 구해 필라델피아의 한 식물원(바트람)에 보냈다. 프랭클린은 '콩을 심을 때 무엇을 조심해야 하는지' 알려주는 편지도 동봉하며 "콩을 갈아 간수로 굳히면 치즈 같은 하얀 식품이

되는데 그것을 두부(tofu)라고 한다"는 것도 알려주었다. 미국인들은 콩과 두부를 들여온 사람으로 벤저민 프랭클린을 꼽는다.

박지원(1737~1805) : 고추장을 담그다

「연암선생 서간첩」은 박지원이 안의현감으로 재임했던 시절 썼던 편지글 32통을 묶은 책이다. 편지의 대부분은 아들들, 특히 큰아들에게 보내졌다. 그는 일찍 부인과 사별하고 혼자 살림을 꾸려가면서 고추장을 손수 담기도 했다. "나는 고을 일을 하는 틈틈이 한가로울 때면 수시로 글을 짓거나 혹 법첩을 놓고 글씨를 쓰기도 하거늘 너희들은 해가 다 가도록 무슨 일을 하느냐? 나는 4년간 「강목綱目」을 골똘히 봤다...고추장 작은 단지 하나를 보내니 사랑방에 두고 밥 먹을 때마다 먹으면 좋을 게다. 내가 손수 담근 건데 아직 푹 익지는 않았다."는 내용이다.

정약용(1762~1836) : 연포회

정약용이 쓴 어원연구서인 「아언각비雅言覺非」 중, 제1권 두부조에는 "모든 능원에는 승원이 있어 두부를 바치는데 이름하여 조포사라 하였다"라는 내용이 있다. 두부가 사찰에서 전수되어진 것은 두부가 소식소찬(素食素饌)으로서 영양학적인 면에서 그 역할이 컸기 때문일 것이다. 또 정약용의 「다산시문집」에는 모두 모여서 연포회를 즐기는 모습이 실감나게 적혀있다.

다섯 집에서 닭 한 마리씩을 추렴하고/콩 갈아 두부 만들어 바구니에 담아라
주사위처럼 두부 끊으니 네모가 반듯한데/띠싹을 꿰어라 긴 손가락 길이만 하게
뽕나무 버섯 소나무 버섯을 섞어 넣고/호초와 석이를 넣어 향기롭게 무치어라
중은 살생을 경계해 손대려고 않는지라/젊은이들이 소매 걷고 친히 고기를 썰어
다리 없는 솥에 담고 장작불을 지피니/거품이 높고 낮게 수나히 끓어 오르네

사무엘 보웬(1760) : 중국의 콩 미국에 전하다

미국의 사무엘 보웬은 1760년 동인도회사의 선원 자격으로 중국을 다녀왔다. 중국과 영국과의 정치적인 이유로 중국 각지를 돌아다니면서 5년간 억류생활을 했는데, 그 때 중국인들이 어떻게 콩을 먹는지 유심히 보아두었던 것 같다. 그는 1765년 콩 한 자루를 메고 그의 고향인 조지아주 사바나에 돌아왔다. 그는 곧 콩을 심었고 이듬해에는 간장을 제조해 영국으로 수출을 시도했다. 그는 간장제조법에 대한 특허를 영국에서 받았지만 건강상 이유로 오래 유지하지는 못했다.

추사 김정희(1786~1856) : 대팽두부와 진장

조선 오백년 역사 동안 가장 뛰어난 학자라는 평을 받고 있는 추사도 콩과 연관이 있다. 그가 쓴 <대팽두부(大烹豆腐)>를 보면 그가 인생의 행복을 소박한데서 찾았음을 알 수 있다.

좋은 반찬은 두부 오이 생강나물(大烹豆腐瓜薑菜)
훌륭한 모임은 부부와 아들딸(高會夫妻兒女孫)

한편 추사가 제주도에 귀양 갔을 때의 일이다. 그는 아내에게 편지를 썼다. “서울서 내려온 장맛이 다 소금 꽃이 피어 쓰고 비위를 면치 못하오니 하루하루가 민망합니다. 경향의 장이 어찌되었는지 빠른 인편을 얻어 내려 보내어야 견디겠습니다. 서울서 진장(陳醬)을 살 도리가 있으면 다소간 사 보내게 하여 주십시오. 변변치 아니한 진장은 얻어 보내도 부질없습니다...”

페리제독(1794~1858) : 일본의 콩을 수집하다.

1850년 일본의 한 난파선이 샌프란시스코에서 구조된 일이 있었다. 일본 선원들은 감사의 표시로 한 봉지의 콩을 선물했다. 이때의 콩 종자는 뉴욕 주 농업협회, 메사추세츠 원예협회, 특허국 등으로 보내어졌다. 1852년 페리제독이 이끄는 중무장한 함대가 일본의 개방을 촉구하기 위해 출동했다. 이때 농업전문가도 탑승했는데 약 1,500~2,000점의 각종 종자와 식물을 수집했다. 그 후 1898년 미국에는 종자식물도입과가 신설됐고 1928년까지 약 30년간 대략 3,000점의 콩 샘플을 일본, 중국, 한국, 인도 등지에서 수집하였다고 한다.

바이에링크(1851~1931) : 뿌리혹박테리아의 발견

오래전부터 콩을 심어왔던 동양에서는 콩을 심으면 땅이 비옥해진다는 것을 진작부터 알고 있었다. 하지만 무엇이 그렇게 만드는지는 몰랐다. 1765년 미국에 처음 콩이 심겨지고 1786년에 독일식물원에, 1790년에 파리식물원에 재배되는 등 당시 서양에서는 콩에 대한 다양한 시험재배가 이루어지고 있었다. 1888년 미생물학자였던 네덜란드의 M.W. 바이에링크는 콩과식물의 뿌리혹에서 박테리아를 순수 분리하는데 성공하였다. 뿌리혹박테리아는 호기성이어서 토양 표층부에 많이 착생한다는 것을 알게 되었다. 또한 뿌리혹박테리아의 종류는 이것이 기생하는 콩과식물의 종류에 따라 다르다는 것도 알게 되었다.

윌리엄 모스(1884~1959) : 미국 콩의 아버지

1929년 미국 농무부에 근무하던 윌리엄 모스는 콩의 다양한 유전자원을 찾아 '동양의 식물 탐험대(일명 콩 원정대)'를 꾸려 동양에 왔다. 콩 원정대가 일본, 한국, 중국 등에서 보낸 총 2년 중 조선에서 보낸 시간은 두 달이 채 안 된다. 그가 수입한 4,578점의 콩 유전자 중 우리나라에서 수집한 콩은 3,379점(73.8%)이었다. 윌리암 모스팀이 길지 않은 시간에 수천 점의 콩 종자를 수집할 수 있었던 것은 우리나라 마을마다 주기적으로 열리는 장터에서 쉽게 다양한 콩을 구할 수 있었기 때문이라는 말도 있다. 윌리엄 모스는 1907년부터 1949년 콩이 미국농업의 중심작물로 성장하기까지 주도적인 역할을 담당하여 '미국 콩의 아버지'가 되었다. 그는 미국대두협회 회장을 3번이나 역임했다.

헨리 포드(1863~1947) : 콩의 왕

'20세기 최고의 경영자', '자동차의 왕'이라고 불리는 헨리 포드의 또 다른 별명은 '콩의 왕'이다. 1929년 미국은 제1차 세계대전과 경제대공항의 여파로 차세대 식량이나 연료에 대한 대안을 찾고자 하였다. 헨리 포드 연구소는 미시간 주에서 실제 경작되는 모든 농작물에 대한 테스트를 실시하였는데, 그때 가장 가치 있는 농산물로 콩이 선택되었다. 콩은 단백질과 지질이 풍부할 뿐 아니라, 섬유질의 용도가 다양하고, 수분이 적어 저장에도 용이하다는 점, 그리고 매년 경작할 수 있다는 점이 높이 평가되었다. 헨리 포드는 78세 생일날, 콩 섬유로 만든 양복을 입었고 콩으로 만든 넥타이를 착용했다. 그는 이어 콩 플라스틱으로 이른바 '콩 자동차'를 만들었다. 차 내부는 콩 플라스틱으로, 차체를 칠하는 페인트는 콩기름에서 얻었고, 연료도 콩기름을 사용했다.

정재원(1917~) : 두유산업의 선구자

정재원(정식품 명예회장)은 콩물에 불과했던 두유를 세계적인 식품산업으로 바꿔 놓았다. 1937년부터 그가 소아과 의사로 재직했던 시절에는 모유와 우유를 소화하지 못하고, 심지어 합병증을 일으켜 사망하는 유아들이 많았다. 1964년 정확한 원리와 병명이 밝혀졌는데 우유 안에 들어 있는 유당을 소화 못하는 '유당 불내증' 때문이었다. 정 회장은 두유가 우유의 대안이 될 수 있다고 판단했다. 그는 2년 만에 발명 특허를 받고 영양식품 허가를 따냈다. 1968년에는 병원 빈터 옆에 수공업 공장을 만들어 생산량을 늘려나갔다. 1973년에는 경기도 용인시에 신갈공장을 만들고 대량 생산을 시작했다. 제품 이름은 '식물(vegetable)'과 '우유(milk)'의 영문을 합성해 '베지밀(vegemil)'로 정했다.

정두화(1918~2010) : 전통장의 부활

정두화옹은 1970년부터 사라져가는 장독대를 복원하고 수백 개의 장독을 두고 규모화를 시작했다. '우리 민족의 정신은 전통장을 지키는 것'이란 기치를 내걸고 시작한 일이었다. 정두화옹은 '큰머슴'을 자처하며 경기도 양평에 전통된장을 담그는 수진원이라는 농원을 조성했다. 옹은 이렇다 할 산업기반이 취약했던 1950년대 우리나라 제일의 구두약 제조회사를 창업하였으나, 일찍이 아들에게 대표직을 물려주고 척박한 황무지 땅을 개간하여 오로지 그 땅에서 수확한 콩으로만 된장을 담근다는 원칙을 지켜냈다. 오늘날 항아리 몇 백 개, 몇 천 개를 갖추어 놓고 전통장을 담그는 이른바 '된장농원시대'를 연 것은 수진원이 최초다. 그는 적어도 전통된장은 2년은 되어야 하고, 간장은 3년 이상은 되어야 제대로 된 것이라는 기준을 마련하였다.

권태완(1932~) : 콩박물관 건립의 주역

1979년 한국과학기술연구소(KIST)의 부서장으로 있던 권태완 박사는 독일의 국제협력 프로그램으로 100만불의 지원을 받아 현 RPC의 전신이 된 '쌀 건조시설 프로젝트'를 수행하게 된다. 1990년 이때의 성과를 발표하기 위해 독일 길에 올랐다가 빵 박물관을 견학하는 기회를 갖게 되었다. 그때 권 박사는 독일에는 빵 뿐만 아니라 맥주, 감자, 아스파라거스 등 단일품목으로 박물관이 10여 군데 있다는 것을 알았다. 그는 돌아오는 비행기에서 콩 문화의 종주국인 우리나라에 콩을 테마로 한 박물관이 있었으면 하는 생각을 했다. 권 박사는 1984년 한국콩연구회와 2001년 콩박물관건립추진위원회를 만드는데 앞장섰고, 2014년 마침내 콩세계과학관 건립의 주역이 되었다.

권신한(1931~) : 한반도, 콩원산지설

우리나라 최초의 콩박사라 불리는 권신한 박사는 우리나라 콩의 유전적변이가 다양하고, 야생콩이 광범위하게 분포되어 있고 중간형도 수집된다는 점에서 '우리나라가 콩의 원산지'라는 개념을 구축하고 콩의 종주국임을 1971년 호주의 국제학술대회에서 발표하여 큰 방향을 일으킨 바 있다. 우리의 조상대대로 이어온 재래종 콩 종자의 소멸을 예방하기 위해 재래종 3,000 계통을 수집, 보존 평가하여 유전자은행을 설치하고 특성을 조사, 분석, 평가한 공로로 1984년도 서울시 문화상 자연과학부문 수상을 하였다. 또한 1998년에는 권태완, 홍은희, 이철호 등과 함께 한국콩연구회지에 <국제규모의 콩박물관 건립에 관한 타당성조사연구>를 진행하여 콩세계과학박물관이 탄생되는데 기초를 마련하였다. 권박사는 미국 농무

부에서 파견된 윌리암 모스팀이 1929~1931년까지 우리나라 중국, 만주, 일본 등지에서 3,000여 계통을 수집해간 일과 1901~1976년 사이에 수집해 간 재래종이 5,496 계통이고 그중 현재 일리노이 대학에 3,200 계통이 보존되고 있음을 밝혀내었으며, 원자력을 이용한 콩의 육종연구를 우리나라에서 최초로 시도하였다.

홍은희(1934~2008) : 콩 교잡육종의 태두

숙정(菽井) 홍은희박사는 우리나라 현대콩을 비롯한 팥, 녹두 등 두류육종과 재배 분야의 태두로써 존경을 받고 있다. 1969년 우리나라 최초의 콩 교잡육성종인 '광교'를 비롯 봉의, 강림, 덕유콩, 황금콩, 장엽콩 등 많은 콩품종을 육성하였다. 숙정은 <한국에 있어서 만파대두의 생육특성과 수량해석에 관한 연구> 등 134편의 연구논문과 7편의 책을 저술하였다. 1986년에는 녹조근정훈장을 받음으로 전 생애에 걸친 콩에 대한 업적을 보상받았다. 1995년에는 정년퇴임기념으로 출간된 「우리나라 두류의 품종」은 당시 우리나라의 주요 두류작물의 전체 품종들을 사진과 함께 실어 이 분야에서 기념비적인 책으로 평가받고 있다.

권순영(1947~) : 아프가니스탄에 콩을 심다

글로벌 식품기업인 네슬레의 영양식품개발담당이었던 그는 우연히 아프가니스탄을 찾았다가 산모와 영유아의 사망률이 굉장히 높다는 것을 알았다. 1차적인 원인은 영양불량이었다. 그는 급한대로 아이들에게 두유를 먹였다. 아이들은 석 달도 안 되어 급격하게 영양이 호전되었고 영유아 사망률도 현저히 떨어졌다. 권박사는 NEI(국제영양기구)를 만들어 좀 더 조직적이고 체계적인 지원을

하고자 하였다. 그는 아프가니스탄 사람들에게 직접 콩을 심도록 권유하였다. 양귀비를 심던 마을에 콩을 심기 시작한지 7년 만에 아프가니스탄의 34개주 전 지역으로 콩재배지역이 확산되는 쾌거를 거두었다. 지금도 아프가니스탄에서 생산된 콩으로 두유공장을 운영하며, 아프가니스탄의 영양문제 해결에 힘쓰고 있다.

세첼(Setchell)(1949~) : 콩의 생리기능성 연구

신시네티 의과대학 교수인 케네스 세첼 박사는 스테로이드 호르몬을 전공하였고 콜레스테롤로부터 담즙산 합성과정 중 유전적 결함을 가진 케이스들을 발견한 것으로 유명하다. 이 과정에서 콩의 이소플라본에 관심을 가지게 되었고 콩의 섭취가 호르몬 관련 질병의 위험도를 낮추는 것을 집중적으로 연구하게 되었다. 그는 이소플라본 분야에 있어 최고 권위자이다. 한국과도 인연이 깊어 1998년 11월 한국콩연구회와 정식품의 공동 주최로 서울 신라호텔에서 열린 '국제콩심포지엄'에 참석하여 "아시아인이 서양인에 비해 질병 위험도가 낮은 것은 생리활성물질이 많이 들어있는 식물성식품을 주로 섭취하기 때문이다"라며 그중 가장 대표적인 것이 콩의 이소플라본이라고 하였다.

Chapter 02

콩과 장(醬)의 문화

1. 콩식품의 기원
2. 전통의 콩 가공식품
3. 우리나라 장 문화

1. 콩 식품의 기원

장(醬)의 기원

문헌상으로 메주를 나타내는 시(豉:메주 또는 두장豆醬)는 중국 한나라(漢, BC 206~AD 208) 시대의 급취편(急就篇)에 처음 나온다. 시를 만들어 큰 부자가 된 사람의 이야기다. 콩이 남만주에서 중국 제(齊)나라로 전해진 것이 기원전 7세기경이라고 하니 한(漢)나라 시대에 두장이 중국사회에서 일반화된 것과 시기적으로 맞는다.[1] 중국의 고문헌들은 시(豉)를 외국에서 유래된 방언으로 보고 있으며, 특히 『신당서新唐書』에는 발해의 특산물로 기록되어 있는 것이 눈에 띈다. 『삼국지三國志 위지魏志 동이전東夷傳』 고구려조에 보면 '고구려 사람들은 장 발효에 뛰어나다(高句麗人 善醬釀也)'라고 기술되어 있는 것을 보면 이 시대 중국인들이 볼 때 고구려는 높은 발효기술을 가진 선진국이었을지도 모른다. 이 무렵, 메주냄새를 고려취(高麗臭)라고 하여 '고려사람 냄새'라 하였다. 오늘날 우리가

---

1) 이철호, 권태완, 콩의 이용역사, 콩, 한국콩박물관건립추진위원회편, 고려대학교출판부, 3-44(2005)

유럽인에게서 치즈냄새를 느끼는 것과 같은 것이다. 즉 치즈냄새가 생소하여 우리가 그 냄새를 맡는 것과 같이 중국인에게 메주나 간장 된장 냄새는 생소하였던 것으로 보이며, 고구려인들은 두장을 주로 먹었으므로 그들의 몸에서 장(메주)냄새를 맡을 수 있었던 것으로 짐작된다.

오늘날 일반적으로 두장을 나타내는 장(醬)은 중국 고전에서 오랫동안 육장(肉醬)을 의미하였다고 한다. 조류나 짐승을 사냥하여 소금에 절인 것을 장으로 표현한 것이다. 육장이 점점 쇠퇴하고 콩으로 만든 간장, 된장이 널리 쓰이면서 장(醬)은 두장을 의미하게 된다.[1)]

『삼국사기』에 보면 신문왕(638)이 김흠운의 딸을 부인으로 맞이할 때 납채품목으로 보낸 물품에 쌀, 술, 기름, 꿀, 육포, 젓갈 그리고 장(醬)과 시(豉) 등이 135수레라고 한 기록이 있다. 고려 숙종(1103)때는 송나라 사람인 손목은 『계림유사鷄林類事』에서 '장(醬)은 곧 밀조(密祖)'라 하였다. 밀조는 메주를 말하는데 시대에 따라 며조, 메조, 미주로 불리었다. 일본의 『정창원문서正倉院文書(739)』와 10세기초 『화명초和名抄』에서도 고구려장이 일본에 와 미소라고 하였다. 1717년에 [동아東亞]에서도 "고려의 장인 말장이 일본에 건너와 고구려 방언 그대로 미소라고 한다"라고 하였다. 동양에서 장(醬)이라는 글자가 처음 나타낸 것은 『주례』 1세기경인데 "해(醢)와 혜(醯)를 말한다"고 했다. AD 100년경에 나온 『설문해자』에서도 '장은 해(醢젓갈)이며 고기와 익은 술로써 만든 장'이라고 하였다. 한(漢)대 이후 동아시아에서 두장은 점차 시(豉)와 말장(末醬), 그리고 곡류를 넣은 다양한 장으로 분화, 발전되었다. 일본의 학자(川田正夫)는 고대로부터 중국에서는 육장(肉醬)이, 일본의 경우는 어장(魚醬)이,

1) 이성우, 한국식문화사, 교문사, 1984

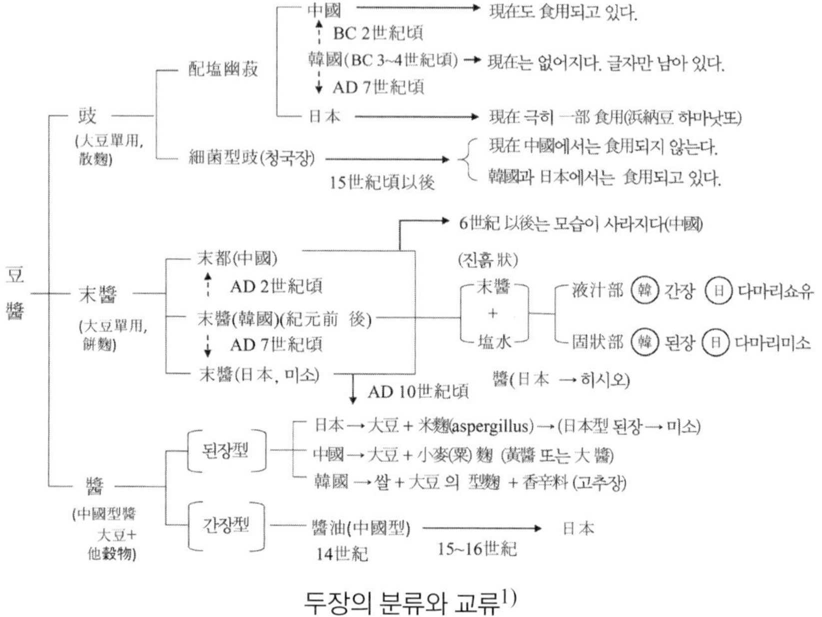

두장의 분류와 교류[1]

조선에서는 두장(豆醬)이 주류를 이루어왔다고 정의하였다.

우리나라에서 언제부터 장이 만들어졌는지는 확실치 않으나 삼국사기에 장의 기록이 보이고 일본의 700년대 초의 문헌에도 고려장이란 기록이 남아 있어, 적어도 1400년 이상의 역사기록이 있음을 확인할 수 있다. 조선 명종 9년(1554)에는 우리나라 최초의 장 기술서 라고 볼 수 있는 『구황촬요救荒撮要』가 이택(李澤)에 의해서 출간되었다. 하지만 그보다 먼저 세종의 어의였던 전순의(全循義)가 『산가요록山家要錄』[2]에서 말장, 합장, 간장 등 13가지의 장 담기를 소개하고 있다.

---

1) 이성우, 고대 동아시아 속의 두장에 관한 발상과 교류에 관한 연구, 한국식문화학회지, 5(3), 313-316(1990)

2) 1450년경 어의(御醫) 전순의(全循義)가 지은 가장 오래된 음식책으로 229가지의 조리법이 나옴

조선시대 초엽의 장류 및 관련 낱말

사시찬요초(四時纂要抄)	장(醬), 포장(泡醬), 즙저(汁菹)
단종실록(端宗實錄)	진장(陳醬)
훈몽자회(訓蒙字會)	장(醬), 첨장(甛醬), 됸쟝, 장유(醬油), 시(豉), 전국시(戰國豉), 두시(豆豉, 전국장)
구황촬요(救荒撮要)	포장(泡醬), 태각장(太殼醬), 태엽장(太葉醬), 도실장(桃實醬), 청장(淸醬, 곤장), 말장(末醬, 메죄, 메조)
미암일기초(眉巖日記草)	간장(艮醬), 감장(甘醬), 태장(太醬), 말장(末醬), 즙저(汁菹)
쇄미록(鎖尾錄)	감장(甘醬), 간장(艮醬), 즙장(汁醬), 포장(泡醬), 비지장(比之醬), 난장(卵醬), 말장(末醬)
동의보감(東醫寶鑑)	장(醬), 두장(豆醬), 소맥장(小麥醬), 육장(肉醬), 어장(魚醬), 시(豉, 된장류, 전국장), 황증(黃蒸 날메주)

장의 어원

우리나라 고유의 간장과 된장은 콩과 소금을 주원료로 하여 콩을 삶아 이것을 띄워 메주를 만들고, 메주를 소금물에 담구어 발효시킨 후의 이 여액을 간장이라 하고, 나머지 찌꺼기를 된장이라 하여 식용해왔다. 간장의 '간'은 소금의 '짠맛(salty)'을 의미하고, 된장의 '된'은 '되직한(hard)'의 뜻이 있다. 간장은 『규합총서閨閤叢書(1809년)』에 '지령'이라 표기되어 있고 서울말로 '지렁'이라 하였다. 이 어원은 아직 밝혀지지 않았으나, 『훈몽자회訓蒙字會(1527년)』의 고어인 '간쟝'과 함께 사용되어 온 말이다. 간장은 단백질과 아미노산이 풍부한 콩으로 만들어지는 발효식품으로 불교의 보급과 더불어 육류의 사용이 금지됨으로써 발생되었고 더욱 발전되었다고 보고 있다.[1]

콩나물의 기원

콩을 재배하기 시작한 기원전 2000년경의 동북아사람들은 물만 닿으면 싹이 자라나는 콩나물의 존재를 일찍 알게 되었을 개연성이 크다. 하지만 콩나물을 언제부터 어떻게 식품으로 먹게 되었는

1) 이철호, 권태완, 한국식품학입문, 고려대학교출판부(2003)

지 정확히 밝혀진 바가 없다. 콩나물은 고려 고종 때 『향약구급방 1236』에 대두황권(大豆黃卷)[1]이 나오는데 약으로 소개되고 있다. 이는 원대(元代)에 나온 『거가필용居家必用』에 녹두나물이 두아채(豆芽菜)라고 소개되고 있는 것보다 빠른 시기다. 우리나라 말로는 '콩을 싹을 내어 기른다'는 뜻으로 '콩기름'이란 호칭으로 불리어 왔다. 이는 지금도 사용되는 엿기름(보리싹 내어 말린 것)이란 단어에서 콩기름과의 유사성을 찾아볼 수 있다. 콩기름이란 호칭은 조선조 말엽 19세기 중반까지도 사용되었다. 장지현[2]에 따르면 우리나라의 경우 두부제조 이전에 콩나물 재배를 하였을 가능성이 있어 콩나물의 재배가 삼국시대로 소급될 수 있다고 하였다.

두유의 기원

맷돌의 원시형인 연석이 한반도의 신석기시대 중기 및 후기 유적에서 발견되는 것으로 보아 두유(豆乳)의 역사도 대단히 오래된 것으로 본다. 두부가 제조되기 이전에 두유가 먼저 식용으로 사용되었을 것이라는 것은 짐작된다. 두유의 호칭은 우리 고유의 낱말로는 '콩국', 한자로는 '두즙'이 있다. 우리나라에서 두즙 문화의 개화는 두부 문화와 같은 시기에 시작되었고, '두부즙' 또는 '두즙'이라는 호칭은 고려시대까지 계속 이어졌다. 조선조에 들어서서 대두를 '태(太)'라고 하기 시작하면서 '태즙' '태포즙'이란 용어가 사용되었으며 조선조 후기에 들어서면서 '두유'라는 낱말이 공존하기 시작하여 오늘에 이르렀다.[3]

1) 생콩으로 콩나물(길금싹)을 길러 햇볕에 말린 것
2) 장지현, 한국 전래 대두이용 음식의 조리·가공사적 연구, 수학사(1993)
3) 손헌수, 두유·두부의 제조 역사와 현황, 콩, 한국콩박물관건립추진위원회편, 고려대학교출판부, 313-364(2005)

두부의 기원

두부는 오래전부터 한국, 중국, 일본 등지에서 제조하여 섭취해 왔던 단백질 이용 제품으로 우리 식단에서도 중요한 위치를 차지하여 왔다. 학계에서는 락(酪요구르트), 치즈 등 유목민들의 식생활에 익숙했던 동이족들이 정착생활을 시작하게 되면서 두부를 처음 만들었을 가능성이 크다고 보고 있다. 중국에서는 한나라 회남왕(淮南王) 유안(劉安)이 저술(BC 178~122)한 『만필술萬筆術』을 근거로 유안이 두부 시조라는 것을 정설로 받아들이고 있다. 이러한 추론은 두부제조공정으로 보이는 그림이 한대의 무덤 마왕퇴에서 발견된 것에 근거를 두고 있다. 그러나 이 벽화는 두부제조공정이 아니라 술 제조공정이라고 추정하는 이도 있다.[1)]

중국의 『청이록淸異錄(960)』에는 처음 두부가 소개되는데 "검소한 생활을 강조하여 고기 대신에 두부를 먹기를 권장하였다"는 대목이 나온 다는 것을 근거로 두부는 당대 말엽부터 만들었을 것으로 식품학계에서는 보고 있다.

우리나라에 두부가 전래된 시기는 분명치 않지만 육당 최남선은 『조선상식』에서 당나라(618~907년) 때라고 하였다. 두부에 관한 최초의 기록은 고려말기 성리학자 이색(李穡)의 문집인 『목은집』의 '대사구두부내향(大舍求豆腐來餉)'이라는 시(詩)에서 발견할 수 있다. '나물국 오래 먹어 맛을 못 느껴 두부가 새로운 맛을 돋우어 주네. 이 없는 사람 먹기 좋고 늙은 몸 양생에 더없이 알맞다.'라는 내용이다.

---

1) 최덕경, 대두의 기원과 醬, 豉 및 두부의 보급에 대한 재검토, 역사민속학 제30호, 한국역사민속학회, 363-427(2009)

2. 전통의 콩 가공식품

『성호사설(1763경)』의 대두론을 보면, "콩(菽숙)은 오곡의 하나인데 사람들이 귀하게 여기지 않는다. 그러나 곡식으로 사람을 살리는 것으로 주장을 삼는다면 콩의 힘이 가장 큰 것이며, 가난한 백성이 얻어먹고 목숨을 잇는 것은 오직 이 콩뿐"이란 말이 있다. 집 주위에 논두렁 밭두렁에 콩을 심고 콩밥, 두부, 콩나물, 장을 담아 먹는 것은 민간에서 할 수 있는 가장 쉽게 영양식품을 얻을 수 있는 방법이었을 것이다.

두 부

두부가 역사와 전통을 가지고 우리 음식에 오랫동안 전래되어 온 것은 두부가 지닌 고유의 담백한 맛과 조직 특성 때문이라고 할 수 있다. 두부를 만들려면 먼저 질 좋은 국산 콩을 골라 콩을 불린다. 여름철에는 7~8시간, 겨울철에는 24시간 정도 불리면 충분하다. 잘 불려진 콩을 맷돌에 간다. 이때 콩 불린 물을 조금씩 부으며 갈아주는데 콩과 물의 분량은 2대 3의 정도가 적당하다. 콩이 갈아지면 눋지 않도록 계속 저으며 끓인다. 끓여진 콩물은 무명천에 받쳐 콩물을 걸러낸다. 콩을 갈아 먼저 걸러내어 콩물(두유)과 비지를 구분하고 콩물만 끓이기도 한다. 여기에 간수를 넣고 저어주면 서서히 응고된다. 이때 물컹한 것이 순두부다. 간수는 소금 가마니를 괴어놓은 후, 가마니 밑으로 떨어지는 물을 받아 사용한다. 작은 구멍이 뚫린 네모난 틀 안에 천을 깔고 갓 엉긴 순두부를 넣은 다음 천으로 싸서 뚜껑을 덮은 후 무거운 돌을 얹어 15분에서 20분 정도 물기를 빼면 맛있는 두부가 완성된다. 이쯤에서 가끔 손으로 살며시 눌러보면서 시간을 조절하면 더욱 부드럽고 맛있는 두부를 만들 수 있다.

콩나물

콩 자체에는 비타민 C가 없으나 발아되면 콩나물이 되어 많은 양의 비타민 C가 생성된다. 성호 이익은 “가난한 자는 콩을 갈고 콩나물을 썰어 합쳐서 죽을 만들어 먹기도 하였는데 족히 배를 채울 수 있다.”고 한 바 있다. 콩나물은 콩나물무침, 콩나물국 등 우리 민족의 오랜 전통음식이자 중요한 식품으로 이용되어 왔다. 콩나물을 집에서 키우려면 먼저 물이 잘 빠지는 용기를 준비한다. 그 다음에는 콩을 준비하여 콩이 완전히 잠기고 남을 만큼의 충분한 물에 3~4시간 담가둔다. 물에 담가두는 시간이 지나치게 오래 두면 싹트는 활력이 떨어지므로 너무 오래 담그는 것은 좋지 않다. 그 다음에는 1일 5~6회 나눠 물을 준다. 물은 자라는 콩나물 몸체에 붙어 있는 각종 유기물들이 잘 씻겨 내려갈 정도로 충분히 준다. 물의 온도가 높으면 빨리 자라지만 부패하기 쉽고, 물 온도가 낮으면 생장 속도가 늦어지므로 미지근한 정도로 준다. 콩나물을 키울 때는 반드시 빛이 없는 어두운 공간에서 키워야 하는데, 밀폐하지 말고 검은 천을 씌우는 것이 좋다. 콩나물은 재배 시작 후 5~6일이면 먹을 수 있다.

된 장

된장과 간장은 음식의 간을 맞추고 맛을 내는 기본식품으로 장 담그기는 일년 중 김장과 함께 가장 중요한 행사로 여겨왔다. 장에 쓸 메주는 보통 음력 10월~12월에 콩을 삶아서 만들어 띄우며 이듬해 입춘 전, 추위가 덜 풀린 이른 봄에 장을 담근다. 소금물에 깨끗이 손질한 메주를 넣어 장을 담그고 사흘쯤은 장독 뚜껑을 덮어 두었다가 햇볕이 좋은 날 아침에 뚜껑을 열어 하루 종일 볕을

쬐어주고 저녁에 덮는다. 항아리 아가리는 면포나 망사로 씌워 이물질이 들어가지 않게 한다. 장을 담근 지 보통 40~60일 정도의 숙성 기간이 지나면 메주와 즙액을 분리한다. 즙액은 간장이 되고 메주부분은 된장이 된다. 장 가르기는 장 담근 시기와 지방에 따라 조금씩 차이가 있다. 장을 가를 때에는 메주가 부서지지 않도록 잘 건져내고 항아리 바닥에 남은 메주 부스러기는 체로 받쳐서 건진다. 건져 낸 메주는 다시 소금을 넣고 버무려 다른 항아리에 꾹꾹 눌러 담는다. 장맛은 메주와 염도, 볕쬐기, 숙성 정도에 의해 결정된다. 소금물의 농도가 너무 낮으면 숙성과정이나 보관 중에 변질될 우려가 있고, 너무 짜면 미생물의 발효가 억제되어 장맛이 떨어진다. 메주 양에 비해 소금물이 많으면 간장의 양이 많아지고 맛은 옅어지며, 물이 적으면 간장의 양은 적고 맛이 진하다. 따라서, 맛있는 장을 담그려면 물은 적게 붓고, 메주를 많이 넣으면 된다. 보통 메주콩 : 소금 : 물의 비율은 1 : 1 : 3~4로 한다.[1,2]

된장의 종류

- **막된장** : 간장을 빼고 난 부산물, 우리가 흔히 '된장'이라고 부르는 장이다.
- **토장** : 메주만으로 담은 된장으로 간장을 빼지 않고 상온에서 장기 숙성
- **막장** : 날메주를 가루로 빻아 소금물로 질척하게 말아 익히는 장으로, 중부 이북에서는 담그지 않고 강원도와 경상도에서 특히 잘 담근다. 충청도에서는 보리밥에 메줏가루, 고춧가루를 섞고 소금으로 간을 하여 담그고, 경상도에서는 콩과 멥쌀을 섞어 만든 메주로 담근다.

---

1) 한복려·한복진, 한국인의 장, 교문사, 2013
2) 신동화, 이효지, 콩 발효음식, 콩, 한국콩박물관건립추진위원회편, 고려대학교출판부, 365-406(2005)

- **담뿍장** : 메주를 곱게 빻아서 고춧가루를 섞고 물에 풀어서 하룻밤 동안 재웠다가 간장과 소금으로 간을 한다. 충청도에서는 메줏가루와 고춧가루를 두부의 순물에 풀고, 황해도에서는 보리밥을 죽처럼 쑤어서 메줏가루, 고춧가루, 소금을 섞어서 만든다.
- **빰장** : 된장만을 목적으로 메주를 굵직하게 빻아 소금물을 끓여서 식힌 물로 담근 장인데 경상도에서 담근다.
- **빠개장** : 메줏가루에 콩 삶은 물, 고춧가루 및 소금을 섞어서 담근 장으로 충청도에서 담근다.
- **가루장** : 보리쌀을 갈아 찐 것에 메줏가루를 버무려, 끓여 식힌 소금물을 부어 간을 맞춘 장으로 강원도에서 담근다.
- **보리장** : 보리쌀을 삶아 띄운 다음 가루로 빻은 것과 메줏가루를 반반씩 섞어서 담근 장으로 제주도에서 담근다.
- **청태장** : 마르지 않은 생콩을 시루에 삶고 쪄서 떡 모양으로 만들어 콩잎을 덮어서 띄움 청태콩 메주를 뜨거운 장소에서 띄워 햇고추를 섞어 간을 맞춤. 콩잎을 덮는 이유는 균주가 붙어서 분해를 용이하기 위함
- **팥장** : 팥을 삶아 뭉쳐 띄워 콩에 섞어 담근다.

오래 끓여야 된장이 맛있는 이유

메주를 잘 띄운 장이라도 단백질이 완전히 분해되지는 않고 펩타이드의 형태로 많이 남아 있다. 따라서 이것을 뚝배기에 넣고 끓이면 끓일수록 펩타이드가 아미노산까지 분해되어 맛이 좋아진다. 그러나 개량메주로 끓인 된장찌개는 오랫동안 끓이면 아미노산이 채소, 고기, 두부 등에 스며들고 국물은 담백해진다. 그러므로 개량메주로 담근 시중된장은 짧게 끓여야 맛이 좋고 전통방법으로 만든 된장은 오래 끓여야 제 맛이 난다.

간장

『증보산림경제1766년』에서는 '장醬은 장수 장獎이니 모든 맛의 으뜸'이라고 하였다. 장은 여러 음식에 간을 하고 맛을 내는 것이므로 음식 중에 제일로 치고, 그래서 때를 잃지 않고 담가야 한다는 것이다. 또한 민가의 장맛이 좋지 않으면 아무리 좋은 채소나 맛있는 고기가 있어도 좋은 요리가 될 수 없고, 반면 시골 사람이 고기를 쉽게 얻지 못하여도 좋은 장이 있으면 반찬에 아무런 걱정이 없다고 하였다. 간장은 음식의 간을 맞추는 기본 양념으로 짠맛, 단맛, 감칠맛 등이 어우러져 독특한 맛과 향이 있다. 간장은 콩으로 메주를 쑤어 소금물에 담근 뒤에 그 즙액을 달여서 만든 장이다. 간장은 달여도 되고 그대로 숙성시키기도 한다. 보통은 된장과 간장을 가르고 나면 간장은 따로 불에 달이는데, 이것은 간장이 부패하는 것을 막고 농축시켜 맛이 진한 장을 얻기 위해서이다. 간장은 80℃의 온도에서 10~20분 정도 지속해서 달이고, 달이면서 생기는 거품은 걷어낸다. 달인 장은 완전히 식힌 후에 독에 붓고 뚜껑을 덮는다. 묵은 간장에 햇간장을 부어 겹장을 만드는데 쓰기도 한다.

간장의 종류

농도에 따라 진간장, 중간장, 묽은 간장으로 나눌 수 있다. 각각 짠맛, 단맛의 정도와 빛깔이 다르므로 음식에 따라 쓰이는 용도가 각기 다르다. 햇간장은 담근 햇수가 1, 2년 정도 되는 것으로 국을 끓이는 데 쓰고, 중간장은 3, 4년된 간장으로 찌개나 나물을 무치는 데 쓰고, 진간장은 담근 햇수가 5년 이상 되어 색이 진해 약식, 전복초 등을 만드는 데 이용했다.

선조때의 합장사

예전에 국난을 당해 임금님이 피난을 가게 되면 피난지에 먼저 가서 임금이 드실 장을 마련해놓는 관직으로 합장사(合醬使)가 있었다. 선조가 정유재란을 당해 함경도에 신(申)씨 성을 가진 관리를 합장사로 선임했는데 조정의 대신들이 모두 반대하고 나섰다. 그 이유가 바로 장을 담글 때 피하는 날인 '신(辛)일'과 '신'이라는 음이 같아서 장맛을 버릴 염려가 있다는 것이었다. 그래서 신씨 성을 가진 가문에서는 사돈네 집이나 딸네 집에 가서 장을 담가 가져오기도 했다. 장맛을 얼마나 중요하게 여겼는지 알 수 있는 일화다.

고추장

고추장은 찹쌀이나 엿기름에서 오는 단맛, 소금의 짠맛, 메주의 감칠맛, 고추의 매운맛이 조화를 이룬 소스로 중국이나 일본에는 없는 우리만의 독특한 장류다. 된장이나 간장이 삼국시대 이전부터 있어 왔다면 고추장은 임진왜란 무렵 고추가 유입되면서 우리 음식에 새롭게 편입되었다. 처음엔 된장에 고춧가루가 조금씩 첨가되다가 차츰 현재의 고추장의 모습이 되었으니 고추장, 고추와 장(醬)의 만남은 조선시대 최고의 퓨전식품이라 여겨진다. 고추의 이용 전과 이용 후가 확연히 구분될 정도로 고추를 이용한 김치, 고추장, 찌개류 등의 반찬들은 우리 밥상을 훨씬 다채롭게 했다. 고추장은 담글 때 섞는 전분질에 따라 찹쌀고추장, 멥쌀고추장, 보리고추장 등으로 구분된다. 고추장용 메주는 콩만으로 만드는 간장메주와는 달리 찹쌀가루를 콩의 20% 정도 넣어 만든다. 고추장은 고추장 메줏가루에 질게 지은 밥이나 떡가루, 또는 되게 쑨 죽을 버무리고 고춧가루와 소금을 섞어서 간을 맞춘 후 발효시킨 조미식품으로 비빔밥은 물론 찌개, 매운탕, 생채, 조림, 회, 강회의 양념 등에 쓰인다. 특히, 생선의 비린내를 없애주므로 생선조림이나 찌개에서는 필수적인 양념이다. 약고추장과 같이 고기를 넣고 볶은 것은 밑반찬으로도 애용된다.

고추장의 종류

- **보리고추장** : 충청도 지방에서 많이 담그는데, 보리쌀을 깨끗이 씻어 가루로 빻아 시루에 찐다. 이것에 끓여서 식힌 물을 섞고 다시 시루에 넣어 더운 방에 놓고 띄운다. 하얗게 곰팡이 폈을 때 고춧가루와 메줏가루를 섞고 소금으로 간을 하여 항아리에 담는다. 분량은 보리 2말에 고추 10근 정도이며, 엿기름을 쓰지 않는 것이 특징이다.
- **수수고추장** : 소금물과 수수가루로 죽을 쑤고 여기에 메줏가루, 엿기름가루, 고춧가루를 섞고 소금으로 간을 맞추어 담근다.
- **무거리고추장** : 메줏가루를 만들고 남은 무거리와 보릿가루, 엿기름가루, 고춧가루를 섞어 담그는 것으로 주로 찌개 고추장으로 쓰는데 맛이 새큼하고 달다.
- **약고추장** : 고기를 곱게 다져 갖은 양념을 하여 번철에 기름을 두르고 볶다가 고추장, 파, 생강, 설탕을 넣고 볶아 만든다. 식은 뒤에 잣을 섞으면 더욱 좋다.
- **팥고추장** : 멥쌀을 흰무리 지고, 콩과 팥은 푹 삶아 절구에서 응어리가 없도록 찧어 반대기를 만들어 위와 같은 방법으로 고추장을 담근다.
- **고구마고추장** : 삶은 고구마에 엿기름을 넣어 삭힌 것을 삼베자루에 넣어 짜서, 이 물을 엿 딜이듯이 졸여서 고춧가루, 메줏가루, 소금을 넣고 만든다.

이성계와 고추장

전라북도 순창 만일사에는 고려말 이성계가 스승인 무학대사가 기거하고 있던 구림면 만일사를 찾아가는 도중, 어느 농가에 들러 고추장에 점심을 맛있게 먹었는데 후에 그 맛을 잊지 못해 왕이 된 후 진상토록 했다는 일화가 소개되어 있다. 이때 이성계가 먹은 것은 조피나루 열매로 만든 초시(椒豉)라고 한다. 『소문사설』에는 순창고초장제법이, 『규합총서閨閤叢書1815』에서는 순창고추장이 지방의 특산물로 나오는 것을 보았을 때 순창고추장이 예로부터 유명했음을 알 수 있다. 순창군은 1997년 고추장 마을을 조성한 이래 장류연구소, 장류박물관, 장류체험관 등 점차 '장류(醬類)밸리'로서의 면모를 갖추고 있다.

청국장

청국장의 유래에 대해서는 여러 가지 설이 있으나 두장류의 가장 초기적인 형태인 시(豉)에서 유래되었다는 설과 17세기 병자호란 때 청나라 군대의 식량으로 쓰이던 속성장이 유입되어 이때부터 청국장(또는 戰國醬)이라고 부르게 되었다는 설이 있다. 전쟁 중이나 유사시에는 장이 익을 때가지 오래 기다릴 수 없어 바로 만들어 먹을 수 있는 속성장이 필요했을 것이다. 흔히 청국장을 병자호란 때 청나라 병사들이 삶은 콩을 말에 싣고 다닌 데서 유래되었다고 말하지만 이성우 박사는 “수천 년 동안 장 문화를 유지해온 조선이 수렵문화를 배경으로 일어나 여진족의 청나라에게서 청국장을 배웠다는 것은 어불성설’이라고 주장한다. 된장과 간장이 메주로 장을 담가 2가지로 나눠졌다면 청국장은 콩이 가진 원래의 물질과 발효산물을 그대로 가지고 있는 효율성이 높은 단백질식품이 된다. 요즘 시중에 많이 나오는 배양균을 첨가하면 하루 만에도 만들 수 있다. 자연발효에 의한 청국장은 메주콩을 10~20시간 더운 물에 불렸다가 물을 붓고 푹 끓여 익힌 다음 보온만으로 띄운 것이다. 시골에서는 소쿠리에 삶은 콩을 한소끔(60℃)까지 식힌 다음 면포를 덮고, 따뜻한 곳에 놓고 담요나 이불을 씌워 2~3일 정도 보온하면 바실러스균이 번식하여 발효물질로 변한다. 바실러스균은 40~45℃에서 잘 자라며, 발암물질을 감소시키고 유해물질을 흡착해서 몸 밖으로 배설시킨다. 바실러스균은 공기 중에도 많이 있지만 볏짚에 많이 들어 있으므로 청국장을 띄울 때 콩 사이사이에 볏짚을 넣고 띄우면 매우 잘 뜬다.

청국장의 종류

- **생청국장** : 청국장을 띄워 생으로 먹는 것이다. 김이나 김치에 싸서 먹기도 하고 비빔밥의 주재료로 이용하기도 한다.
- **청국장환** : 생청국장을 건조시켜 가루로 내었다가 다시 먹기 좋게 환으로 제조한다. 청국장만을 환으로 만들기도 하고 찹쌀가루를 섞거나 솔잎가루, 다시마가루 등을 넣어 만들기도 한다.
- **청국장가루** : 생청국장을 건조하여 가루로 만든다. 청국장가루는 요구르트에 타서 함께 먹거나 우유, 생식 등에 타서 먹기도 한다.
- **말린청국장** : 생청국장을 그대로 건조시킨 것이다. 땅콩처럼 바삭바삭한 맛으로 먹는다.

3. 우리나라의 장(醬)문화

택일과 장 관리

예로부터 장은 모든 음식의 기본이라 여겼기 때문에 좋은 장을 얻기 위해 많은 노력을 기울였다. 말날을 택해 장을 담그는 등 최대한 부정타지 않도록 조심했고, 장담기 사흘 전부터 나들이를 삼가고 장담는 날에는 심지어 음기가 발산될까봐 입을 한지로 봉하기도 하였다. 장독대를 관리하는데도 지성을 다 바쳤다. 항아리나 단지들은 언제나 먼지 하나없이 반질거리게 닦아놓으며, 장항아리는 햇볕이 나면 열고 비가 오려고 하면 서둘러 닫아 빗물이 한방울이라도 섞이지 않도록 조심했다. 만일 장맛이 변하기라도 하면 그것은 주부의 큰 실책일 뿐 아니라 집안에 변고가 닥칠 징조라 믿었기 때문이다. 손님을 초대할 때는 "오셔서 우리집 장맛이나 보시지요"라는 인사말을 하기도 했다.

메주를 지푸라기로 묶는 이유

메주를 만들 때 지푸라기로 묶는 것은 지푸라기에 붙어 있는 세균 고초균(*Bacillus subtilis*)이 메주에 잘 접종되도록 하려는 것이다. 고초균은 임의적 혐기성균이며 고온을 좋아하는 (최적온도 40도씨 부근)내열성균이다. 메주는 공기와 접촉하는 외벽에 곰팡이들이 자라고 내부에 고초균을 포함한 세균들이 자란다. 메주 발효온도는 26~30℃부근이 적당하고 온도를 너무 높이거나 자외선이 강하거나 수분이 부족하면 좋은 메주를 만들 수 없다.

숯과 고추를 넣는 이유

장을 담글 때 숯을 넣는 이유는 장맛을 변하게 하는 잡귀를 숯 구멍에 가둔다는 주술적인 의미도 있지만 실제로 숯에는 수많은 나노 사이즈의 미세한 구멍이 있어 자연 속의 유익한 미생물이 거기에 자리를 잡아 장이 잘 발효되도록 도와주는 역할을 한다. 고추는 고추의 붉은 색과 매운 맛으로 장맛을 변하게 하는 잡귀를 멀리 쫓는다는 의미가 있지만 실제로 고추에 들어있는 캡사이신 성분이 살균과 방부효과가 있어 장맛이 변질되는 것을 막아준다. 대추의 붉은 색도 잡귀를 물리친다는 의미로 넣지만 장을 달게 하는 역할을 한다.

버선목 거꾸로 붙이기

장항아리에 한지로 만든 버선본을 거꾸로 붙여 놓거나 줄을 매달아 두기도 하는데, 이는 장을 해치는 귀신이 버선 속에 들어가서 나가지 못하게 한다는 의미였다. 즉 장맛이 상하지 않게 하려는 주술적인 의미가 있다. 한편 과학적인 근거도 있는데, 다족류 벌레들은 흰 한지가 반사하는 빛을 싫어하기 때문에 항아리 입구에

흰 버선을 붙여 놓으면 벌레들이 얼씬거리지 않게 된다고도 한다. 충남 논산지방에서는 버선본을 거꾸로 붙이고 나서 "꿀독이오"라고 외친다. 꿀처럼 맛있는 장이 되어 달라는 뜻이다.

숨 쉬는 옹기

장을 담는 옹기는 장독이라 해서 특별히 골랐다. 장독은 이른바 물을 통과시키고 공기는 통과시키는 이른바, 숨 쉬는 옹기를 써야 한다. 옹기는 오뉴월에 구운 독은 습기가 완전히 제거되지 않아서 좋지 않으며, 겨울에 구운 독을 이른 봄에 사야 좋다고 하였다. 두드려보아 쇳소리 나는 독이 좋은 독이다. 예로부터 옹기는 '숨 쉬는 그릇'으로 인식되어 왔다. 찰흙에 들어있는 수많은 모래 알갱이가 그릇 벽에 미세한 공기구멍을 만들어 옹기의 안과 밖으로 공기가 통하기 때문에 안에 담긴 음식물이 잘 익고 오래 보존된다. 이런 장점 때문에 옹기는 발효를 위한 최고의 용기라 할 수 있다.

각 지역별 옹기의 특성

옹기는 지역에 따라 각각 독특한 형태를 이루고 있다. 장독대는 각종 옹기를 질서정연하게 큰 것(뒤쪽)에서부터 작은 것(앞쪽)으로 경사를 형성하도록 배열한다. 전체적으로 보면 따뜻한 남부지방의 옹기는 항아리 통이 넓고 폭을 넓게 하여 빛의 투과율을 낮췄으며 북부지방은 옹기 폭을 좁게 하여 빛을 많이 받아들이도록 되어 있다. 서울·경기도 지방은 입과 밑지름의 크기가 비슷하여 날씬한 모습이다. 강원

도 지방은 지역적으로 인접한 경기도 지방과 비슷한 골격을 이루고 있으며 어깨부분의 경사가 급하고, 입이 넓은 형태도 존재하며 산악 지역에서 운반하기 쉽게 작은 형태의 옹기도 보인다. 충청도 지방은 목이 높고 밖으로 약간 벌어진 형태이다. 전라도 지방은 입이 밑바닥에 비해 더 넓고 윗배가 불렀으며, 어깨에서 밑으로 둥글게 흘러내려 떡 벌어진 모습을 하고 있다. 경상도 지방은 몸통부분이 유난히 돌출되었고, 어깨에서 입까지 급격하게 좁아져 입이 매우 작은 모습이다. 제주도 지방은 입과 밑이 좁고, 배가 부른 형태이며 화산토로 만들어 붉은색을 띠고 있다. 그러나 다양했던 옹기의 지역적 특성은 지역간 인적 물적 교류가 활발해지면서 점차 희미해졌다.

속담 속의 콩과 장

콩이나 장과 관련된 속담은 80여개에 이른다는 조사가 있다. ‘마음은 콩밭에 있다’ ‘콩나물 시루같다’ ‘가뭄에 콩 나듯’ ‘사랑을 하면 눈에 콩깍지가 씐다’ ‘먹을 콩을 알고 덤빈다’ ‘콩 심은 데 콩나고 팥심은데 팥난다’ ‘꿩이 콩밭 생각하듯’ ‘두부에서 뼈라’ ‘두부먹다 이 빠진다’ 등은 콩과 관련된 속담이다. ‘한 고을 정치는 술맛으로 알고, 한 집안 일은 장맛으로 안다’는 말도 있고 ‘며느리가 잘 들어오면 장맛도 좋아진다’ 또 ‘흥하는 집은 장맛도 달다’고까지 하였으니 아녀자들에게 장 담그는 일은 겨울나기의 중요한 행사일 수밖에 없었던 것이다. 이밖에 들어보면 무슨 말인지 짐작이 가는 재미난 속담도 많다. ‘집장을 십년하면 호랑이도 안 먹는다’. ‘말 단 집에 장이 곤다’ ‘말끝에 장 달란다’. ‘된장 맛으로 이불 속의 며느리를 들춰본다’ ‘고추장이 밥보다 많다’ ‘고린 장이 더디 없어진다’ ‘개에게 된장 덩어리 지키게 하는 격’ ‘간장국에 마르다’ 내손에 장을 지진다‘ ’간장에 전 놈이 초장에 죽으랴’

4. 다양한 콩 음식

우리조상들은 콩을 재배하여 콩을 이용한 음식을 일상식은 물론, 의례음식으로도 이용하였다. 다음은 이효지 교수에 의해 조사된 콩을 이용한 음식 분류이다.[1)]

콩을 이용한 음식 분류

▪ 주식류

밥에는 콩과 콩나물, 죽에는 콩, 콩나물, 두부, 고추장, 된장, 국수에는 콩, 콩가루 콩물, 두부, 만두에는 만두 껍질에 두부, 만두소에 두부, 범벅에는 콩, 떡국에는 두부가 이용되었고, 식량사정이 어려울 때 삶은 콩을 식용하기도 했다.

▪ 부식류

국(탕)에 콩, 두부, 콩가루, 콩나물, 된장, 비지, 조치(찌개)에는 콩비지, 두부, 순두부, 된장, 청국장, 고추장, 콩나물, 지짐이에는 콩나물, 두부, 고추장, 전(부침, 지짐, 각두포, 저냐)에는 두부, 나물(콩나물, 콩나물볶음, 숙아채, 콩나물 무침)에는 콩나물, 콩자반(두자반, 콩장)에는 콩, 조림(두부장전, 장자, 두부조리개, 두부장, 두부장조림)에는 두부, 구이(적)에는 두부, 생채에는 두부, 장아찌, 두부, 콩잎(단풍든 콩잎), 회에는 두부, 쌈에는 콩잎, 김치에는 콩나물, 콩잎이 이용되었다.

---

1) 이효지, 우리나라 콩 음식들, 콩, 한국콩박물관건립추진위원회편, 고려대학교출판부, 455-528(2005)

의례음식

콩을 이용한 찐 떡에는 콩을 쌀가루와 섞어서 떡을 찌고 콩가루는 찐 떡의 고물이나 송편소로 이용되었다. 친 떡에는 콩가루가 고물로 이용되었다. 지진 떡에는 된장이 이용되었다. 삶은 떡에는 콩가루가 고물로 이용되었다.

과정류

다식, 강정에는 콩가루가 이용되었고, 산자, 빙사과, 엿강정, 엿에는 콩이 이용되었다.

음청류

콩을 볶아서 끓인 물을 받쳐서 마시거나, 콩을 볶아서 가루로 만들어서 냉수에 타서 꿀이나 설탕을 넣고 마셨다.

1) 콩밥 2) 콩나물밥 3) 콩나물죽 4) 콩죽 5) 콩국수 6) 연포갱 / 7) 비지찌개 8) 순두부찌개 9) 청국장찌개 / 10) 된장찌개 11) 두부선 12) 콩설기 13) 콩찰편 14) 쇠머리떡 15) 오쟁이떡 16) 인절미 17) 부편 18) 된장떡 / 19) 찹쌀경단 20) 다식 21) 콩강정 22) 오향두부간 23) 십경두부 24) 고소방 / 25) 여의두부 26) 두유온노 27) 반죽콩밥 28) 고모구찜 29) 고과두부볶음 30) 두부피자

다양한 콩음식

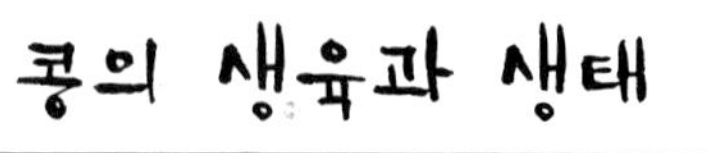

1. 우리나라 콩 품종과 품종 해설

우리 민족은 예로부터 콩을 오곡의 하나로서 쌀과 함께 아주 주요한 먹거리 작물로 여겨 왔다. 우리나라에서 콩은 삼국시대에 이미 메주를 만들어 장류용으로 이용하였고, 통일신라시대 및 고려조에 들어서면서 두부 및 콩나물용으로 그 용도가 확대되었다. 그리고 시작 연대는 정확히 알 수 없겠으나 콩은 우리 식생활에 깊이 파고들어 혼반용, 떡소용, 자반용 등 다양하게 이용되어 왔으며, 콩 품종의 분화도 용도에 맞게 개량되고 발전되어 왔다.[1] 콩 품종에 대해 기술한 것을 농서에서 볼 때 조선조 초기 기록으로 보이며, 저자가 미상이긴 하지만 『사시찬요초四時纂要抄』에는 청태, 흑태, 황태로 콩의 품종을 크게 색깔로 구분한 바 있으며, 성종조에 강희맹이 지은 『금양잡록衿陽雜錄1492』에는 콩 8품종에 대한 품종해설을 해 놓았다.

---

1) 김석동, 이영호, 콩 품종과 육종, 콩. 한국콩박물관건립추진위원회편, 고려대학교출판부, 137-216(2005)

조선조 말 경기도 수원에 권업모범장이 세워진 1906년 이후 재래종 콩의 수집선발이 이루어져 재래종 콩으로서 품종으로 등록된 것은 장단백목을 비롯한 충북백, 부석 등 28품종이나 되었고, 이들은 오늘날 세계 제일의 콩 생산국인 미국 콩의 유전적 기반이 되었다. 또한 이들은 1969년 인공교배를 통하여 최초로 개발된 광교품종이 등장되기 이전 까지는 전국 각 지역의 콩 재배 명품으로 생산 이용되었다.

그 후 2014년까지 농촌진흥청, 각도 농업기술원과 대학에서 개발 등록한 콩 품종은 200여종에 달한다. 이중 두부와 장류용 콩으로 널리 재배되고 애용된 것은 80년대와 90년대 초까지에는 장엽콩과 황금콩이었고, 90년대 중후반부터 현재까지 태광콩, 대원콩 등이 전국을 풍미하고 있다. 한편 콩나물콩으로는 재래종인 준저리, 서목태, 백좀콩과 도입종인 힐콩, 단엽콩 등에 이어서 1980년대 후반에서 1990년대 초반까지는 은하콩이 많이 보급되었고, 오늘날에는 새롭게 개발 보급된 풍산나물콩이 높은 생산성과 기호성으로 널리 사랑받고 있다.

또한 1993년 이후 검정콩1호를 시작으로 선흑콩, 청자콩에 이어서, 청자콩2호, 청자콩3호 등이 자반용 또는 밥밑용 유색콩으로 개발 보급되어 재래종인 서리태와 함께 밥밑용 콩 수요에 대응하게 되었으며 1991년 이후 큰올콩과 화성풋콩, 선녹콩, 미랑 등의 개발 보급으로 올콩과 풋콩 등의 수요에 대응하여 왔다.

80년대 후반부터는 시장개발 드라이브와 WTO체제에 적응할 수 있는 품질 고급화가 크게 요구됨에 이에 부응해서 콩 품종들이 개발되었다. 콩 특유의 비린내는 지질 산화효소인 리폭시게나제(lipoxigenase)에 의하여 발생하는데 이를 제거한 진품콩 시리즈는

두유용이나 콩 아이스크림을 만드는데 적합한 품종이다. 단백질 함량이 45% 이상으로 두부수율이 일반콩보다 15% 이상 높은 고단백질 품종인 단백콩, 새단백콩 등이 개발 생산되고 있다. 또한 최근에 개발된 대풍콩은 80년대 중반 이후에 나온 팔달콩과 그 시리즈인 신팔달콩2호의 초형과 다수성 유전자들을 직간접적으로 받은 품종으로 품질은 다소 떨어지나 콩 재배 안정성과 수량성이 현 개발된 어느 품종보다도 우수한 품종으로 농가에 보급되고 있다.

야생콩의 순화와 재배콩의 육종

야생콩은 콩알이 너무 작거나 혹은 성숙기가 되면 콩깍지가 저절로 열려 콩알이 튀는 등 단점이 있는데, 선사시대의 선조들에 의하여 순화되어 오늘날의 재배콩이 되었다. 재배콩을 개량하기 위하여 재래종의 순계분리, 해외 자원 도입, 교배 등의 육종 과정을 통하여 새롭게 개발된 콩을 신품종이라 하며 국가적으로 재배를 장려하여 널리 보급하게 되었다.

우리나라 콩 육종사업은 1906년 권업모범장이 설립된 후 많은 지방재래콩을 수집하여 비교평가와 순계분리를 하면서 시작되었고, 장단백목은 1913년 순계분리육종법에 의하여 선발된 최초의 장려품종이며, 1969년 최초의 교잡육종 품종인 ‘광교’가 개발 육성되면서 본격화되었다.

1980년 콩알이 굵고 색택이 좋은 ‘황금콩’이 개발되어 우리나라 장류·두부용 콩의 고급화를 선도하였고 1990년대 태광콩, 대원콩, 2000년대 대풍콩, 우람콩 등이 그 뒤를 이으면서 우리나라 주력 콩 품종으로 보급되었다.

콩의 품질을 높이기 위해서는 무엇보다도 각 용도별로 특성을 개발하고 그러한 고유특성을 가진 품종을 개발하여야 한다.

2. 콩의 일생

콩의 일생[1)]

---

1) 한원영, 기능성작물부, 국립식량과학원, 2014

3. 용도별 품종 분류

여기에서 말하는 콩이란, 장(醬)을 담아 먹는 장콩을 말한다. 콩은 용도에 따라 장콩, 메주콩, 콩나물콩, 두부콩 등으로 불리며 색깔에 따라 노란콩, 황두, 검은콩, 흑태 등으로 불리기도 한다. 여기서 콩이란 단백질 함량이 40%에 이르는 대두(大豆)를 말하는 것이다. 강낭콩이나 완두콩은 전분이 60%, 단백질이 12~16%에 달해, 곡류에 더 가깝다고 할 것이다.

콩의 품종

- **백 태** : 황두, 노란콩 또는 흰콩, 백태, 된장 등의 장을 담그는 기본 재료인 메주를 만드는 데 가장 널리 쓰여 '메주콩'이라고도 한다. 백태는 우리나라에서 가장 많이 생산되는 콩이다. 메주로 만들어 된장, 간장, 고추장 등을 담그는 데 가장 많이 쓰인다. 두부를 만들어 먹기도 하고, 싹을 틔워 나물로 먹기도 한다.
- **서리태** : 껍질은 검은색이지만 속이 파랗다고 하여 '속청'이라고도 부른다. 작물의 생육 기간이 길어서 10월경에 서리를 맞은 뒤에나 수확할 수 있으며, 서리를 맞아 가며 익어간다고 하여 '서리태'라고 하였다. 물에 담갔을 때 잘 무르고 당도가 높아 다른 잡곡과 함께 밥에 넣어서 먹거나 떡을 만들 때 함께 넣는 등 주로 식용으로 쓰인다. 서리태를 발효시켜 만든 청국장은 건강 효과가 좋을 뿐 아니라 맛도 메주콩으로 만든 청국장보다 뛰어나다. 안토시아닌 색소가 많아서 검은콩의 대명사로 꼽히는데, 안토시아닌 성분은 꾸준히 장기 복용하면 노화를 방지하는 효과가 있다. 서리태는 단백질과 식물성 지방질이 매우 풍부하고, 신체의 각종 대사에 반드시 필요한 비타민 B군, 특히 B_1, B_2와 나이아신 성분이 풍부하다.

- **서목태** : 일명 '쥐눈이콩(鼠目太)'. 쥐눈처럼 작고 검은콩이다. 민간에서는 '약콩'으로 불리기도 한다. 유황을 뿌리고 재배하면 약성이 더욱 좋아지는 것으로 알려져 있다. 성질이 따뜻하고 맛은 달며, 독이 없다. 신장과 관련한 곡식이므로 신장병이 있을 때 먹으면 좋으며, 광물성 약재의 독을 비롯한 모든 독을 풀며, 혈액순환을 활발하게 한다. 볶아서 먹으면 몸이 더워지고, 술에 담갔다가 먹으면 풍증(風症)에 효과가 있다. 삶거나 찐 콩에 소금·새앙 따위를 섞어 띄워서 약을 만들면 성질이 몹시 차가워진다. 이것으로 죽을 쑤어 먹으면 소갈증을 없애주며, 장을 만들면 성질이 평범해진다. 그러므로 알맞게 약을 만들어 써야 한다. 쥐눈이콩에는 이소플라본 성분이 일반 콩보다 5~6배 많이 함유되어 있다.

장류용

장류는 콩 발효제품으로서 우리나라에서 가장 대표적인 전통식품 중의 하나이다. 근래에 이르기까지 콩 육종의 근간을 이루어 온 것이 장류용 콩 육성이었다.

장류용 콩의 특징

장류용 콩은 단백질 함량이 높고, 대립[1]으로 종피색과 배꼽색이 황색인 것이 양질로 취급되며 품질개량도 고단백으로 종피 및 배꼽이 황색이면서 대립쪽으로 이루어져 왔다. 이 밖에도 맛, 냄새, 제품의 색깔 및 수율 등 가공적성을 고려하면 장류용 콩의 특성은 대립뿐만 아니라 흡수율 및 무름성도 높아야 되며 종피 두께도 얇아야 하고, 콩 특유의 냄새도 장류용 콩의 가공적성으로 고려해야 한다.

1) 백립중의 무게 : 콩의 무게를 재는 방법으로 콩 100알의 무게를 잰다. 백립중의 무게가 40g 이상은 극대립, 25~40g은 대립, 15~25g은 중립, 10~15g은 소립, 10g 이하의 것은 극소립으로 구분

장류에 좋은 품종

작물시험장(1998)은 우리나라 콩과 수입콩 등 21품종을 공시하여 30여 가지의 품질 특성을 조사하고 메주제품에 대한 맛 관련 성분의 변화양상을 추적하여 고품질 메주가공용으로 적합한 품종을 1차 선발하였다. 메주의 유리아미노산, 포르말태 질소 등 맛 관련 성분의 발효기간별 변화양태를 조사하여 메주용 우수 콩으로 최종적으로 선발된 품종은 예상과는 달리 '단백콩'(14.9g/100립), '두유콩'(20.5g/100립), '단엽콩'(15.3g/100립)으로 중립에 속한 품종이었다. 앞으로 장류용 콩은 지금까지의 외관상 품질에 치중한 선발보다는 장류용으로서 메주나 청국장 등 가공적성에 맞는 품종을 선발하는 방향으로 전환하여야 할 것이다. 그러나 중립 또는 소립의 가공적성이 높은 품종은 수입콩과 차별화가 어렵고 시장성이 낮으므로 재배단지를 만들어 가공 후에 출하하는 것도 이와 같은 문제점을 해결하기 위한 방법이 될 것이다.

두부 및 두유용

두부 가공원료로 사용되는 콩의 품질과 두부의 각 제조공정은 두부의 수율과 조직감, 향미, 겉모양 등 두부 품질에 많은 영향을 준다. 콩의 화학성분 중 지방과 단백질이 두부 및 두유에 가장 잘 옮겨지고 탄수화물 중에서도 섬유는 두유박, 콩깻묵(대두박)에 남고 당질은 응고과정 중에 대부분 압착액으로 나간다.

두부 및 두유용 콩

단백질과 지질 함량이 높은 것이 두부 원료로서 좋은 것이며, 특히 수용성 단백질 함량이 높을수록 두부수율이 높다. 탄수화물로는 소당류인 라피노즈(raffinose)와 스타키오즈(stachyose)의 함량

이 적으며, 영양적으로 필요한 무기질의 흡수를 방해하는 피트산(phytic acid)과 단백질 소화를 억제하는 트립신 저해제의 함량이 적은 콩이 두부의 영양적 품질을 향상시킬 수 있다. 이 밖에도 콩은 두부제조 과정 중 물에 침지한 뒤 마쇄하게 되므로 흡수속도가 비교적 빠른 품종이 두부 제조가 쉽다. 또한 종피에 칼슘과 조섬유의 함량이 높으면 흡수속도가 감소하며 종피의 두께 및 탈피 정도도 두부수율에 영향을 미친다.

두부 및 두유에 좋은 품종

단백질 함량과 두부수율로 볼 때 '백운콩', '단백콩', '장엽콩' 등이 대체로 우수하며, 1993년에 육성한 '단백콩'은 단백질 함량이 45% 이상으로 두부수율이 일반 콩 대비 13%가 높으나 100립중이 14g으로 소립인 것이 단점이다. 1980년대 후반부터는 비린내 없는 콩 품종 육성을 위해 노력한 결과 리폭시게나제(lipoxygenase)가 불활성화된 '진품콩'과 '진품콩2호'를 개발 보급하였다. '진품콩'과 '진품콩2호'는 두유가공 공정에서 콩 비린내를 제거하기 위한 가열처리 과정을 생략할 수 있는 장점이 있으나 기존 두유 가공공정이 가열처리를 하도록 되어 있어서 두유용으로 보급되지 못하는 것이 아쉬운 점이다.[1)]

나물용

콩나물은 비닐하우스에서 채소가 재배되기 이전까지는 김치와 더불어 동절기 중 가장 중요한 비타민 C 공급원이었다. 나물용 콩은 종피의 색, 크기, 무게, 흡수속도 등도 콩나물의 품질에 많은 영향을 끼친다.

---

1) 콩의 비린 맛을 내는 lipoxygenase를 불활성화 시키기 위해서는 가열처리를 해야 한다

나물용 콩의 특징

콩나물은 주로 배축을 식용으로 하기 때문에 배축의 생산량이 곧 콩나물 생산량이 되므로 나물용 콩은 단시일 내에 빨리 자랄 수 있는 소질을 가진 품종의 선발이 요구된다. 또한 원료 콩의 발아력이 중요한데, 일반적으로 소립이 대립보다 콩나물 생산 면에서 현저히 높은 편이므로 나물용 콩의 적성은 소립종이 우수하다. 이 밖에도 콩나물은 씹을 때 질기게 느껴지는 섬유질이 적은 것이 좋지만 다른 한편으로는 쉽게 부러지는 것을 방지하기 위하여 적당량의 섬유질이 요구된다. 콩나물 선택 시 선호되는 은빛 나는 줄기와 황금빛 나는 자엽 등의 색택도 나물용 콩의 특성으로서 요구된다.

나물용에 좋은 콩

우수한 나물용 콩 품종 육성을 위하여 1970년대부터 노력하였으나 1980년대 초반까지는 육성품종의 콩나물 재배적성이 낮아 콩나물 재배업자들은 주로 재래종을 원료 콩으로 사용하여 왔다. 1984년 처음으로 방사콩이 돌연변이 육종법에 의하여 육성보급 되었고, 1986년에는 은하콩이 보급됨에 따라 재래종을 '은하콩'으로 대체하여 재배하는 농가가 점차 증가하였다. 1990년대 초반까지는 재배농가의 50% 이상이 '준저리'와 '오리알태' 같은 재래종을 재배하였는데, 이는 육성품종이 재래종에 비하여 종실이 굵고 콩나물 수율과 품질 면에서 좋지 않았기 때문이다. 그러나 1990년대 후반에 '풍산나물콩' '소명콩' '소원콩', '도레미콩'과 2000년대 이후 '신화', '풍원' '해품' '장기'와 같은 콩나물 재배적성이 우수한 품종들이 대거 육성됨에 따라 현재는 대부분의 농가가 육성품종으로 재배하고 있으며, 특히 대단위 계약재배를 원하는 콩나물 재배업자들은 수량과 병해충에 강하고 순도가 높은 육성품종들을 선호하고 있는 실정이다.

밥밑용

밥밑콩으로 불리는 검정콩을 비롯한 유색콩은 우리나라에서 쌀, 보리, 잡곡 등과 함께 혼합하여 주식으로 이용할 뿐만 아니라 콩자반, 떡속, 제과용 및 약용으로 이용되어 왔다.

밥밑용 콩의 특징

밥밑용으로 이용되는 유색콩은 자엽이 녹색인 것이 시중에서 인기가 높은데, 이는 일반적으로 종실이 굵고 밥에 넣었을 때 잘 무르고 맛이 좋기 때문이다. 밥밑용 콩의 취반특성 및 식미 등의 특성은 여러 가지 요소들이 복합적으로 작용하여 나타나는데 이는 종실의 외형, 향미, 조직 등도 관여한다. 따라서 밥밑용 품종 육성 목표를 유색 대립이면서 총당 함량이 높고 무름성이 좋으며, 색소용출정도가 높은 품종개발에 두게 되었다.

밥밑용으로 좋은 품종

현재까지 육성한 품종으로는 '검정콩1호', '검정콩2호', '일품검정콩'과 검정색 종피에 자엽이 녹색인 '청자콩', '흑청콩', '청자콩2호, 조생서리 등 14품종이 있다. 이와 같이 밥밑용 품종 개발에도 상당한 성과를 거두었으나 이 품종들은 대체적으로 도복에 약하고 밥밑용 적성이 재래종에 비하여 떨어지며, 일부 품종은 성숙기가 지나치게 늦은 문제점이 있다.

풋콩 및 올콩용

풋콩은 옛날부터 우리나라에서 밥밑용이나 떡소용으로 이용되어 왔으나 언제부터 이용되기 시작하였는지는 확실하지 않다. 1990년대에 들면서 풋콩을 상업적으로 재배하여 시장에 출하는 농가가

눈에 띄게 늘어나 최근에는 수백ha에서 재배되고 있는 것으로 추정되나 정확한 통계자료는 없는 실정이다.

풋콩의 특징

풋콩은 독특한 풍미를 가지고 있으며, 단백질, 지방은 물론 비타민 A, C, E(토코페롤)가 풍부하다. 풋콩의 품질 고급화와 관련된 특성을 보면 1) 콩 꼬투리의 크기가 크고, 2) 꼬투리색은 선명한 초록색이면서 종실색은 노란색 또는 초록색이 좋고, 3) 꼬투리의 모용색은 회색, 4) 한 꼬투리에 2~3립의 종실이 들어 있는 비율이 높고 5) 종실에 탄수화물 함량이 높아서 단맛이 나고 6) 종실에 글루탐산(glutamic acid) 등 맛에 기여하는 아미노산의 함량이 높아 좋은 맛을 내야 하며 7) 향기가 좋고 8) 조직감도 고려되어야 한다.

풋콩으로 좋은 품종

1990년대 이후에 육성한 풋콩 품종으로는 도입종인 '화엄풋콩'과 '석량풋콩', 재래종에서 순계분리한 '화성풋콩', 교배 육성한 '신록콩'과 '선녹콩', '단미풋콩', '다진콩' 등이 있다. 최근 남부지방에서 양파, 마늘 등의 재배가 증가함에 따라 이들 후작물로 올콩이 재배되고 있으나 육성된 품종은 '큰올콩', '새올콩', '검정올콩', '다올콩', '검정새올콩', '한올콩', '황금올콩', 참올콩' 등 8품종뿐이다. 한편 올콩용인 '큰올콩'이나 재래 올콩들도 풋콩용으로 이용되기도 한다. 풋콩이나 올콩 품종들은 생태적으로 거의 같은 조생종 품종들이나 이용목적에 따라 구별된다. 현재 올콩에서 가장 문제가 되는 종실관련 병 저항성인 품종의 육성이 시급히 요구된다.

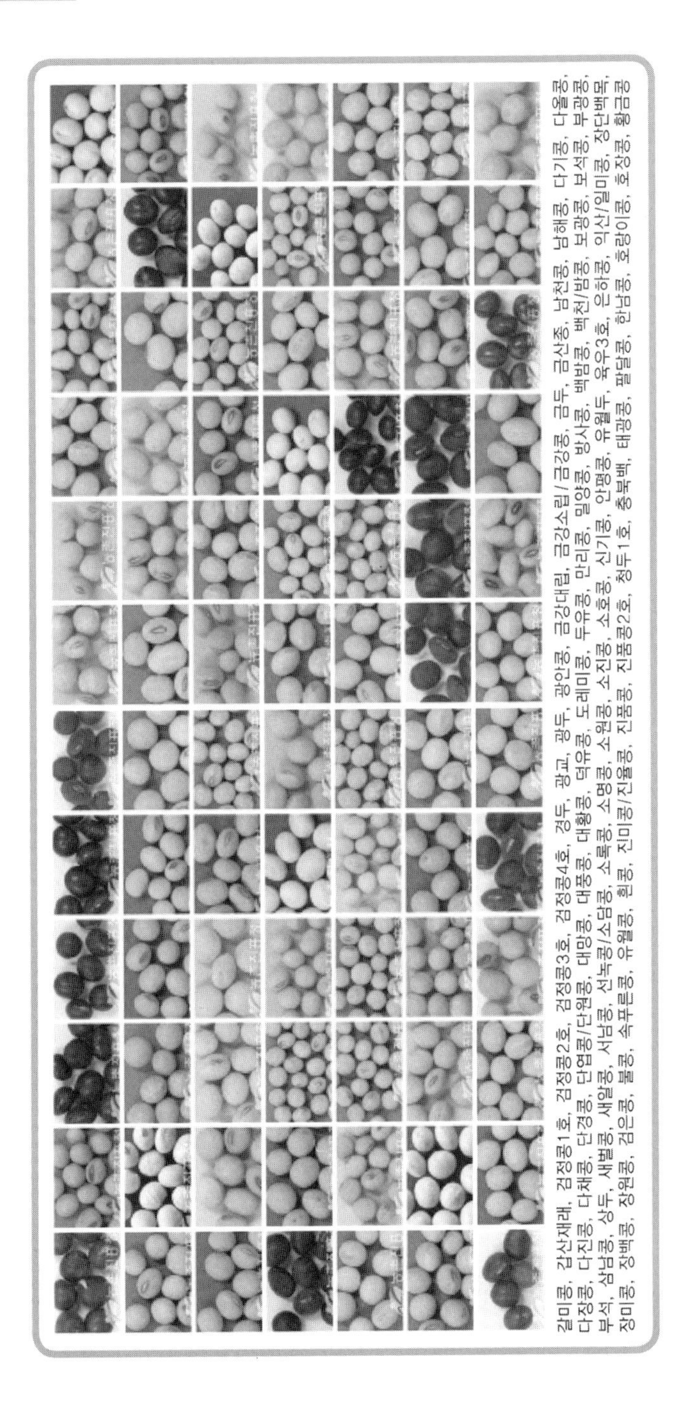

갈미콩, 갑산재래, 검정콩1호, 검정콩2호, 검정콩3호, 검정콩4호, 경두, 광교, 광두, 광안콩, 금강대립, 금강소립/금강콩, 금두, 금산종, 남천콩, 남해콩, 다기콩, 다올콩, 다장콩, 다진콩, 다채콩, 단경콩, 단엽콩/단원콩, 대망콩, 대풍콩, 대황콩, 덕유콩, 도레미콩, 두유콩, 만리콩, 밀양콩, 방사콩, 백밤콩, 백천/밤콩, 보광콩, 보석콩, 부광콩, 부석, 삼남콩, 상두, 새별콩, 새알콩, 서남콩, 선녹콩/소담콩, 소록콩, 소명콩, 소원콩, 소진콩, 소호콩, 신기콩, 안평콩, 유월두, 육우3호, 은하콩, 익산/일미콩, 장단백목, 장미콩, 장백콩, 장원콩, 검은콩, 불콩, 속푸른콩, 유월콩, 흰콩, 진미콩/진율콩, 진품콩, 진품콩2호, 청두1호, 충북백, 태광콩, 팔달콩, 한남콩, 호랑이콩, 호장콩, 황금콩

농업유전자원센터에 보관중인 대표적인 콩

갈미콩, 갑산재래, 검정콩1호, 검정콩2호, 검정콩3호, 검정콩4호, 경두, 광교, 광두, 광안콩, 금강대립, 금강소립/금강콩, 금두, 금산종, 남천콩, 남해콩, 다기콩, 다올콩, 다장콩, 다진콩, 다채콩, 단경콩, 단엽콩/단원콩, 대망콩, 대풍콩, 대황콩, 덕유콩, 도레미콩, 두유콩, 만리콩, 밀양콩, 방사콩, 백밤콩, 백천/밤콩, 보광콩, 보석콩, 부광콩, 부석, 삼남콩, 상두, 새벌콩, 새알콩, 서남콩, 선녹콩/소담콩, 소록콩, 소명콩, 소원콩, 소진콩, 소호콩, 신기콩, 안평콩, 유월두, 육우3호, 은하콩, 익산/일미콩, 장단백목, 장미콩, 장백콩, 장원콩, 검은콩, 불콩, 속푸른콩, 유월콩, 흰콩, 진미콩/진율콩, 진품콩, 진품콩2호, 청두1호, 충북백, 태광콩, 팔달콩, 한남콩, 호랑이콩, 호장콩, 황금콩

농업유전자센터에 보관중인 재래종콩

우리나라는 다른 어떤 나라보다도 콩 자원의 다양성을 보유하고 있으며, 현재 농업유전자원센터는 고유 재래종이 8,132점이 보관되어 있다. 특이한 토종 콩 자원들은 전통식품의 원료로 활용해 왔을 뿐만 아니라 민간의료 소재로 활용해왔기 때문에 건강기능성 측면에서 소비자들의 선호도가 매우 높다. 토종콩 복원과 함께 중요하게 추진되어야 할 부분은 이들 복원품종들을 지역 특산품목으로 육성하는 것이 중요하다.

콩 육성 계보도

콩 육성 계보도는 여러 집안의 족보와 같다. 육성 계보도를 보면 개별 품종의 선조와 다른 품종 간의 유전적 거리등을 추정할 수 있으며 같은 선조종으로 부터 유래하는 품종간의 교배를 가능하면 하지 않도록 하여 유전적 다양성을 유지할 수 있는 참고자료가 된다. 현재 우리나라 품종들은 몇 개 선조품종으로 유래하여 유전적

취약성이 높다. 앞으로는 여기 나온 계보도에 없는 품종과의 교배가 필요하다고 본다.[1]

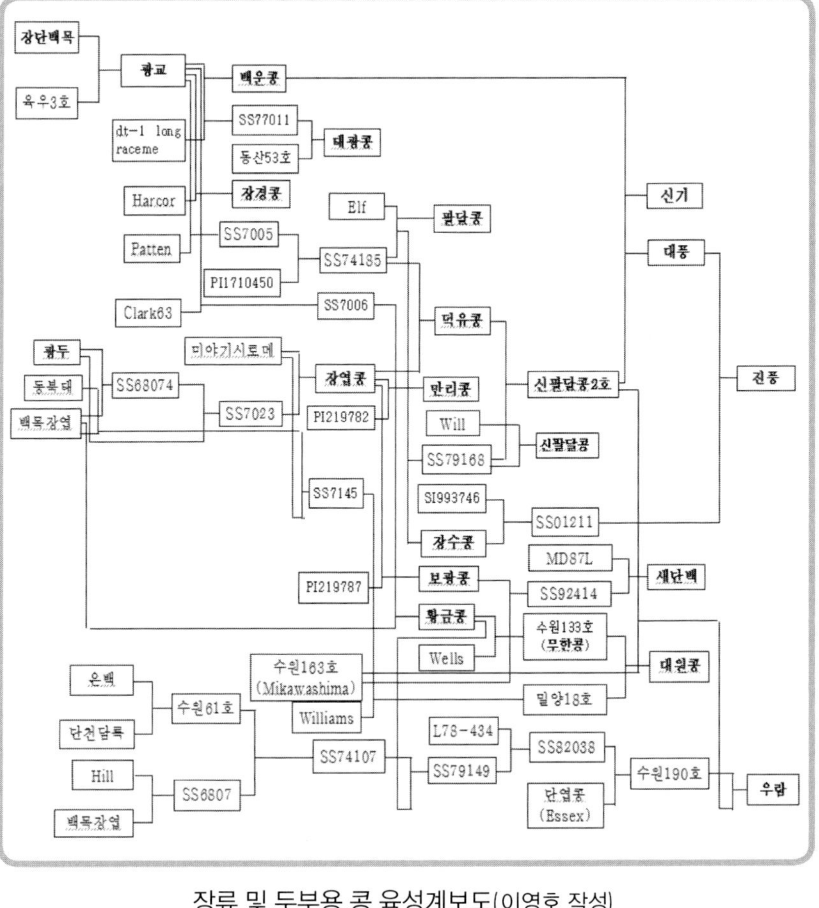

장류 및 두부용 콩 육성계보도(이영호 작성)

1) 김석동, 이영호, 콩 품종과 육종, 콩. 한국콩박물관건립추진위원회편, 고려대학교출판부, 137-216(2005)

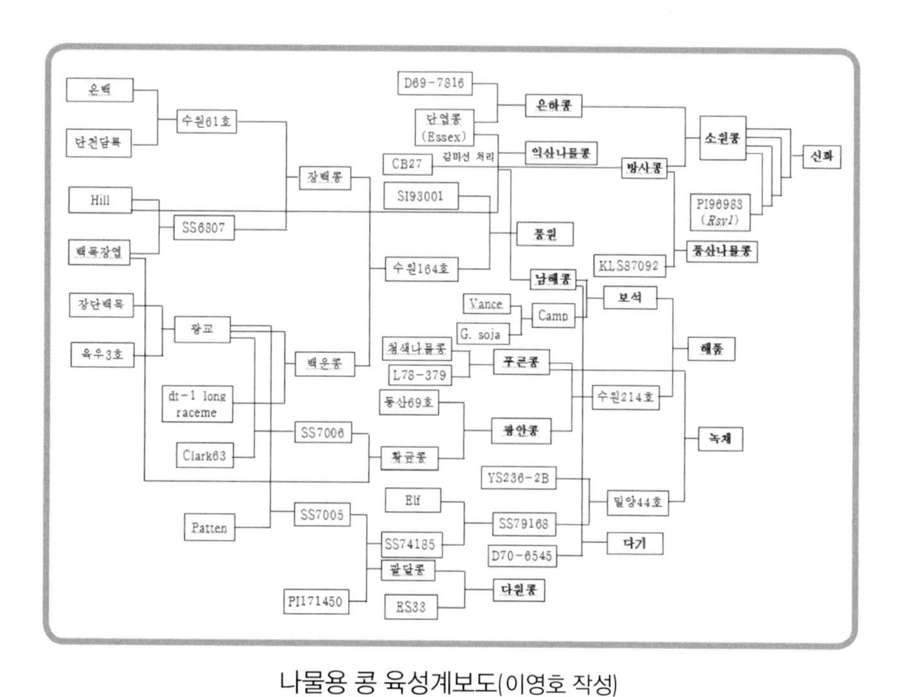

나물용 콩 육성계보도(이영호 작성)

우리나라의 토종콩

우리나라는 콩의 원산지로 콩의 쓰임의 용도가 다양하고 품종과 이름 또한 다양하다. 그 이름들은 대개 지방명을 딴 것, 무늬 또는 색깔을 딴 것이 많고 쓰임새에 따라서 혹은 파종기, 숙기와 열리는 형태에 따라서 이름을 붙이기도 하였다. 또한 우리나라에서는 중북부 지방이나 산간의 콩 단작지대에서 콩이 많이 재배되어 왔으나 남부 지방에서는 알이 작은 품종들이 많이 재배되어 왔다.

색깔과 무늬별

새알콩, 선비밤콩, 청태, 아주까리콩(아주까리밤콩), 흰콩, 누런콩, 검정콩(흑태), 파랑콩, 새파랑콩, 검은밤콩, 밤콩, 자주콩, 속푸

큰콩(속청), 비추콩(비취콩), 푸르대콩, 수박태, 자갈콩, 대추콩, 대추불콩, 쥐눈이콩(서목태), 알종다리콩, 눈까메기콩(눈깜장이콩)

수확시기별

올태, 40일콩, 쉰날거리콩(50일콩), 유월두(유월콩), 서리콩(서리태)

조리용도별

뚝콩, 고물콩, 파란고물콩, 나물콩(지레미콩, 질금콩, 지름콩), 메주콩, 약콩, 콩, 밥밑콩(배미콩, 밥콩, 반미콩)

수집장소에 따라

갑산태, 청산태, 정선콩, 부석, 장단백목, 영양, 울산, 함안, 익산, 단천

형태별

보각다리콩, 준저리콩, 좀콩, 납작콩(납떼기콩, 납드레콩, 납쪼리기콩, 납지르기콩), 한아가리콩, 부채콩(또는 맨드라미콩)

파종장소에 따라

보리밭콩(보리간작용), 논두렁콩(두렁콩)

영주 부석태 1호

영주지역의 재래종 부석태를 교잡 육종법이 아닌, 순계분리 육종법에 의해 육성한 콩이다. 일반 콩 100알의 무게가 20g 내외인데, 부석태 1호는 40g 정도이며 맛과 영양이 뛰어나다. 영주는 오래전부터 콩의 주산지이자 생산지의 명맥을 이어왔다. 1948년에 경북영주지방에서 수집된 재래종 부석은 1960년에 경북지방의 장려품종으로 결정되어 농가에 보급되었다. 영주시의 이러한 노력은 콩세계과학관을 영주시에 유치시키는 계기가 되었고 콩의 문화, 관광, 교육, 산업의 지역 거점화를 목표로 공동연구 및 상호협력을 추구하여 세계적인 콩 연구 및 산업의 메카를 형성하고자 한다. 영주시는 유전적으로 안정화되지 못한 부석태의 단점을 극복하기 위해서 전국 최고의 콩 육종기술을 보유한 농촌진흥청 국립식량과학원과 2009년 4월 양해각서(MOU)를 체결하였다. 그 결과 여러 유전자가 섞여 있는 상태인 부석태를 순계분리법에 의해 개량하였으며 부석태 1호가 신품종으로 출원될 수 있도록 육종기술을 지원하였고 부석태 1호를 재배하는데 적합한 맞춤형 재배기술도 개발하였다.

4. 콩 농사와 재배법

콩의 재배면적은 우리나라 전체 밭 면적 75만 5,000ha 중 약 10% 정도로 아직까지는 밭작물 중 단일작목으로는 가장 많은 재배면적을 차지하고 있으나 콩이 재배되고 있는 밭의 입지조건이 매우 열악한 상태에 있다. 위치별 분포비율을 보면 중, 산간지에 위치한

농가가 조사대상 농가의 69%에 이르고, 이러한 위치에는 기계화가 쉽지 않으며 물주기도 어려워 상습적인 한발의 피해를 입기 쉬운 조건에 있다. 농림축산식품부는 쌀 수급 안정을 위해 논에 벼 대신 콩이나 옥수수를 심도록 해 쌀 생산량은 줄이고 다른 식량작물생산을 늘려 전체 식량자급률을 높이도록 권장하고 있다.

콩 농사

콩을 언제 심는 것이 가장 좋은가 하는 것은 품종의 생태형(조생종, 중생종, 만생종), 기상환경, 재배양식, 작부체계 등에 따라 다르다. 콩 파종은 일반적으로 땅 온도가 15℃ 이상이 되면 가능하다. 콩의 다수확을 위해서는 무엇보다도 파종 후 균일하게 싹이 터서 충분한 발아개체를 확보하는 것이 중요하다. 콩을 파종할 때는 해충에 의한 피해, 출아하지 못하는 경우를 감안해 한 구멍에 2~3알의 콩을 심는다. 우리나라에서 콩 파종이 가능한 기간은 대체로 4월 초순부터 7월 초순까지 약 3개월간으로 보는데 시기에 따라 수량의 차이는 크다. 이와 같이 넓은 파종기를 갖는 것도 콩의 중요한 특성 중 하나이다. 이러한 특성을 이용하여 다른 작물과의 윤작으로 경지 이용도를 높일 수도 있다. 봄철 보리 뒷그루로 콩을 심는 재배방식은 우리나라 대부분의 지역에서 가장 보편적으로 채택되어 온 방식이다. 지방에 따라 옥수수와 혼작(섞어심기), 또는 보리수확 전 골에 미리 콩을 심는 간작(사이심기), 또 보리와 콩의 윤작(돌려심기), 쌀·겉보리와 콩의 윤작 형태 등 다양하다.[1)]

---

1) 홍은희, 콩 재배역사, 콩. 한국콩박물관건립추진위원회편, 고려대학교출판부, 103-136 (2005)

품종의 선택

콩은 용도별로 품종이 개발되어 있음. 된장, 두부, 콩나물, 풋콩 등 이용 목적 및 단작, 이모작 등 작부형태에 따라 성숙기를 고려하여 품종을 결정해야 한다.

파종시기와 방법

적정 파종기는 중부지방은 6월 상순, 남부지방은 6월 상중순. 봄감자, 봄옥수수, 양파, 마늘 등의 뒷그루로 콩을 재배하는 경우에는 파종기가 6월 하순~7월 상순으로 늦어지는데 가능한 파종을 앞당겨야 하며 이때 파종기가 7월 중순까지 늦어지는 경우 콩 수확 후 양파나 마늘 파종기를 고려하여 품종을 선택하여야 한다. 콩을 파종하는 깊이는 3~5cm가 적당한데 너무 깊으면 발아 후 땅 위로 출아하는데 불리하고 너무 얕은 경우에는 토양 수분이 충분하지 않아 발아에 불리하다. 파종장비로는 점파기가 널리 이용되고 있으며 트랙터 부착 또는 인력으로 이용 가능하다.

경운 및 시비

심경은 토양의 공극율, 토양경도, 투수성 등 토양 물리성을 개선하며, 하층토의 지력질소 이용과 심경에 의한 근권 확대로 생육을 왕성하게 하므로 가을이나 이른 봄에 심경을 하는 것이 좋은데 특히, 논토양의 통기성을 좋게 하여 콩의 일생동안 필요한 질소량의 50%를 공급하는 근류균의 증식에 도움이 되며, 건토효과도 얻을 수 있다. 논은 대부분 점질토양이므로 비가 오면 작토층이 굳어져서 파종작업이 어려우므로 파종직전 경운, 정지작업이 반드시 필요하다. 콩은 척박지에서 재배가 가능하다는 잘못된 인식으로 토양개량

없이 콩을 재배하여 왔으나 콩의 질소 흡수량은 고정질소 50%, 지력질소 40% 및 시비질소 10%로 생육하는데 지력질소를 높이려면 토양개량이 필수적이다. 콩의 표준 시비량은 토양의 비옥도가 포장마다 다르므로 토양 검정치에 의하여 시비량을 조절하여야 하는데 최근 농가에서는 대부분 복비를 사용하고 있으므로 이때 질소 요구량을 기준으로 시용하고, 인산과 칼리는 단비로 보충하여 질소 과용에 의한 과번무와 도복을 막을 수 있도록 해야 한다.

콩 수확 후 관리

콩 수확 후 수분 함량은 약 17% 전 후가 되므로 선별과 저장을 위해서 건조시키는 것이 좋다. 처음부터 기계건조 하는 것보다 수확 전 콩밭에서 80% 자연건조 후에 나머지 20%는 건조기 및 그늘에 의해 건조시키는 편이 콩의 품질을 떨어트리지 않으며 선별 및 저장하기에도 좋다. 탈곡된 콩은 종자로 활용할 경우 수분함량이 13% 이하가 되게 잘 건조시키고 다른 품종이나 병충해립이 섞이지 않도록 하며 햇볕에서 건조하는 것이 높은 발아력을 유지할 수 있어 좋고, 기상조건 등으로 건조기 이용이 불가피 할 경우에는 건조 온도를 30~40℃ 내외에서 풍건이 되도록 하여야 한다. 고온에서 급속히 건조하게 되면 종피에 균열이 생기는 등 종자의 품질이 저하될 우려가 있기 때문이다.

콩의 선별

콩 수확 후 건조가 완료되면 선별작업을 한다. 콩의 선별은 대표적으로 기계적 방법인 선별기와 인력 손 선별을 하며, 선별 시 혼종이 되지 않도록 구분관리 해야 한다. 특히 종자용과 판매용

콩을 한 대의 선별기를 이용할 경우 선별 후 청소를 철저히 하여 품종이 섞이지 않도록 주의를 기울여야 하며, 이물질이 섞이지 않도록 하고 크기별로 잘 선별하여야 제값을 받을 수 있다. 콩 재배 농가에서 생산하는 장류용 및 두부용과 나물용 콩은 대부분 기계선별을 통하여 콩의 품질이 표준화 되어 있는 반면 밥밑용 콩은 형태 및 수량 관계로 재배농가들이 인력을 이용한 손 선별을 해야 한다. 콩의 품질향상을 위해서는 기계적 색채 선별을 이용한 철저한 품질관리가 필요할 것으로 전망된다.

콩의 저장

콩 수확 후 저장관리는 생산제품의 품질에 매우 중요한 영향을 미치며, 재배농가에서 보관 기간은 겨울철(12월~2월)이다. 특히 재배농가 대부분 저온 저장시설이 없는 관계로 개방창고에 보관하는데 이때 외기 온도가 상승하면 대사로 인한 단백질 소모가 시작되어 발아세와 영양상태 등 품질이 떨어진다. 콩 저장은 종자의 경우 저장온도, 습도 및 종자의 수분 함량에 따라 수명이 크게 달라지는데 온도와 습도가 높으면 수명이 짧아지므로 잘 말린 종자를 건조하고 서늘한 곳에 보관하도록 하는 것이 좋고 장기 저장 시 해충 피해를 막기 위해 4~7일 정도 훈증제를 처리 보관하는 것도 좋다. 1년 이상 장기저장이 필요한 종자의 경우에는 저장온도 5℃ 이하 상대습도 45%~50%에서 보관해야 한다.

조선시대 초엽의 장류 및 관련 낱말

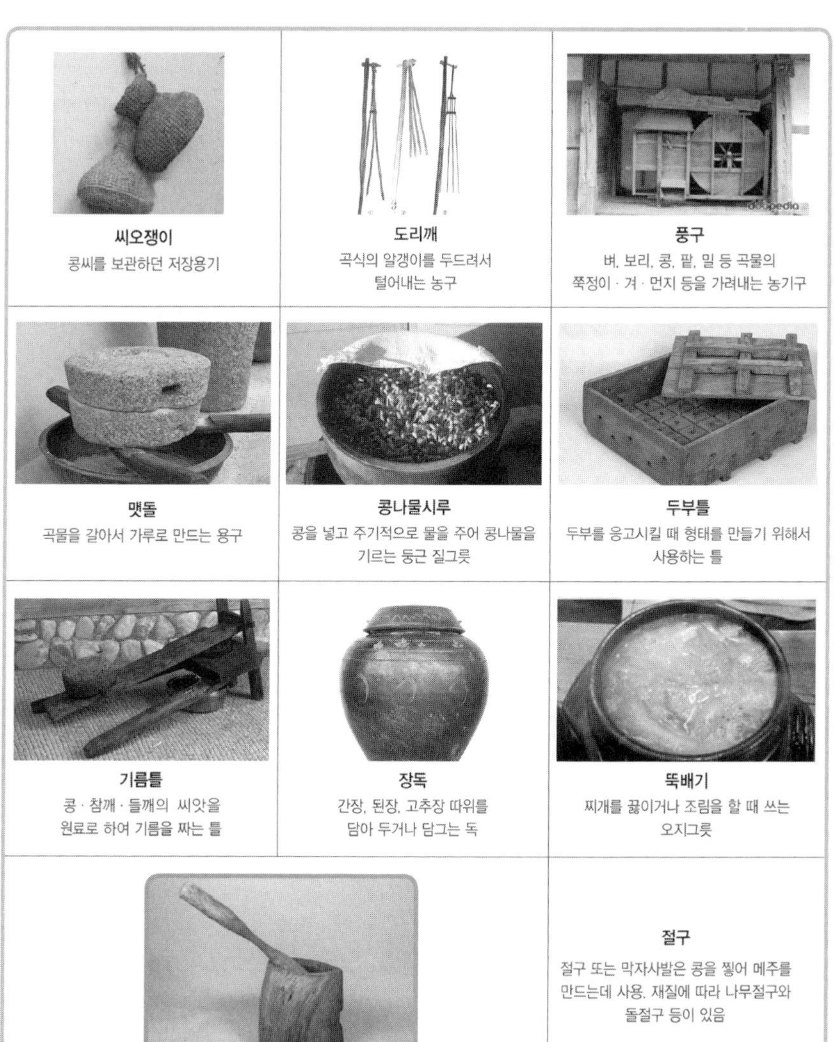

Chapter 04

콩의 가공과 이용

1. 콩의 이용

콩의 가공 및 이용에 관해서는 콩을 오래 전 부터 재배하여 이용해왔던 동북아시아지역이 가장 잘 발달해 있으며 그 이용방법은 역사적, 문화적 배경과 지역에 따라 큰 차이가 있다. 콩을 가공해온 방법을 크게 보면 콩밥이나 콩자반처럼 원형을 변형시키지 않고 조리하는 방법과 마쇄, 발효, 분해, 발아, 추출, 분리건조, 응고, 절임, 조직화 등을 거쳐 원형을 변형시켜 가공한 방법이 있다. 콩의 전체 성분을 이용한 것은 익힌 콩, 된장, 전지콩가루, 전두부 등이 있으며 콩나물은 발아 방법, 두부는 응고방법, 콩기름 및 두유는 추출방법, 콩가루는 마쇄방법, 농축 및 분리 콩단백 등 분말제품은 분리 및 건조방법에 의한 것이다.[1)]

콩의 주요 성분별 이용 제품

콩의 전체성분을 먹는 것은 콩밥의 콩, 콩자반, 된장, 두유, 비지,

1) 김우정, 콩의 가공특성, 콩, 한국콩박물관건립추진위원회편, 고려대학교출판부, 217-312(2005)

전지콩가루 등이고 단백질을 이용한 제품은 간장, 두부류, 유바, 콩치즈, 조직탈지 콩가루 등이다. 지방질을 이용한 제품으로는 콩기름, 콩버터, 토코페롤, 레시틴, 지방산 등이고 콩의 기능성성분을 이용하는 제품에는 이소플라본, 피니톨, 올리고당, 펩타이드 등이 있다.

2. 콩의 위대한 변신

콩을 날 것으로 먹으면 거의 소화가 되지 않기 때문에 가공은 필수적이다. 콩밥으로 먹으면 65% 가량 소화되고 된장, 청국장은 85%, 두부는 95% 정도의 소화 흡수율을 보인다. 단백질은 그 자체로는 맛을 내는 성분이 없으나 펩타이드나 아미노산으로 분해되는 경우에는 감칠맛을 낸다. 전통 메주의 역할은 콩에서 잘 증식될 수 있는 각종 미생물을 자연적으로 끌어들여 증식시킴으로써 이들이 생산하는 효소를 이용하여 콩 단백질을 분해하는 것이다. 따라서 메주의

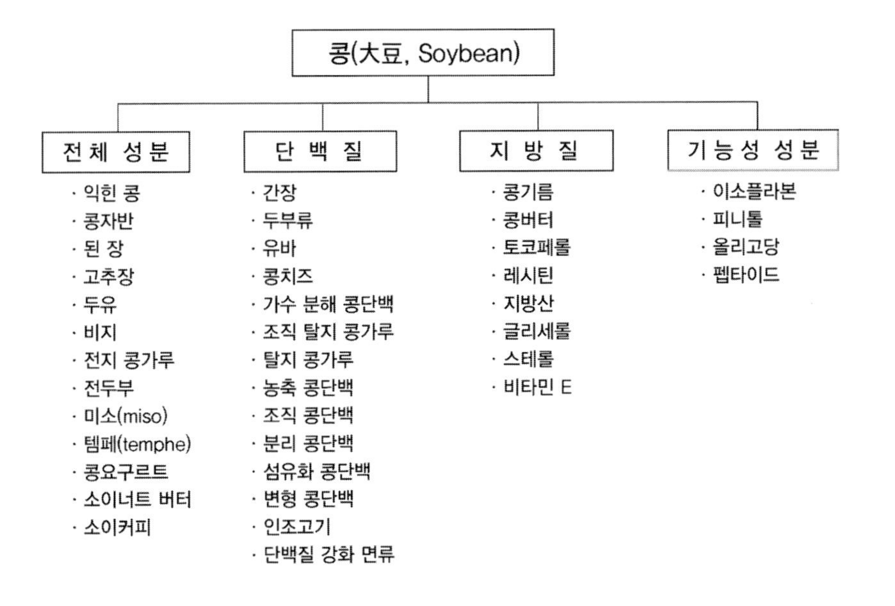

발효경과는 곧 바로 장류의 품질과 밀접한 관계가 있다. 된장, 간장 등 발효식품은 단백질이 많을수록 좋고, 콩기름 제조에는 지방질의 함량과 지방산의 조성이 중요하다. 콩밥에 쓰이는 밥밑용 콩은 물에 침지시켰을 때 수분을 흡수하는 수화속도, 가열할 때의 익힘 속도, 그리고 익힌 콩의 텍스처가 중요하며, 두유나 두부는 수용성 단백질과 수용성 고형분의 양이 주요 품질 요소가 된다. 또한 트립신 저해제, 피트산 등 영양소의 흡수 장해요인 제거와 만성질환 예방에 효과가 큰 기능성 성분의 함유량도 주요 품질요소로 인정되고 있다.

콩 가공식품

▪ 두유 제조

콩을 물에 갈아 그 액을 가열하여 비지를 짜내고 액체를 모은 식품으로 양질의 단백질과 영양성분이 풍부하다. 콩을 잘 씻어 여름에는 7~8시간, 겨울에는 24시간 물에 담가 불렸다가 맷돌에 갈아서 이것을 끓인다. 이 가열로 인하여 콩의 비린내가 제거되는 동시에 단백질이 다량 용출된다. 가열이 끝나면 이것을 베주머니에 넣고 걸러 짜서 콩물과 비지로 나눈다. 이때 콩비지가 너무 식으면 짜기 어려우므로, 뜨거울 때 걸러서 가능한 한 콩물을 꼭 짠다. 얻어진 두유는 음료로, 비지는 찌개나 반찬에 사용된다.

▪ 두부 제조

콩을 물에 충분히 불려 마쇄한 뒤 끓인 다음 이를 여과하여 비지를 제거하고 얻어진 두유에 응고제를 넣어 응고시킨 후 압착한 것이다. 여과 과정 중 불용성 단백질과 고분자 탄수화물 및 상당량의 지방질이 비지로 제거되며 나머지의 지질과 당은 수용성 단백질을 응고시

킬 때 두부 속에 포함되게 된다. 우리나라에서 주로 생산·판매되고 있는 두부는 순두부, 일반 두부, 연두부의 세 종류로서, 가장 많이 섭취하는 일반 두부는 탁한 흰색에 두부 특유의 담백한 맛을 갖고 있으며, 비교적 단단하면서 약간의 탄력성이 있고 조직이 거친 것이 특징이다. 한편 연두부는 그 조직이 균일하며 매끄러운 표면을 갖고 있는 장점이 있으나 조직이 지나치게 연하다. 순두부는 견고성이 연두부보다는 단단하지만 일반 두부보다 약하여 과거의 순두부는 조직이 거칠었으나 최근의 순두부는 균일하게 제조되고 있다. 가공두부류에는 유부, 생양(두부의 표피만 튀긴 것), 군두부, 냉동건조두부, 계란연두부, 분말두부, 두부국수, 강화두부(비타민, 미네랄, DHA 등 첨가), 어육두부, 발효두부, 수프(두부를 발효시켜 보존성을 높인 것), 우유두부, 혼합두부(두유에 야채, 해조류 등을 가열 응고시킨 것), 조미두부 등이 있다.

콩나물 제조

콩나물은 콩을 발아시킨 것으로 가장 전통적인 1차 가공식품으로 콩이 발아되어 성장하는 동안 영양 성분이 크게 달라진다. 종실에 함유된 지질은 크게 감소하는 반면 식이섬유소는 증가하고 특히 비타민 C, B_2, 아스파라긴산(asparagines)이 발아와 함께 급격히 생성되어 채소로서의 가치가 크다. 이외에도 피로회복과 심근경색에 효과가 있는 카르니틴(carnitine), 콩나물 잔뿌리에 많고 알코올 분해효소의 기능을 도와 숙취해소에 좋은 아스파라긴산, 혈압을 낮추는 칼륨, 항암작용이 있는 이소플라본 등이 수침과 발아를 통해 증가한다. 콩나물의 비타민 C 함량은 7일째 최고가 된다. 반면, 숙취해소에 좋은 아스파라긴산의 함량은 기를수록 많아진다.

발효식품

된장 제조

전통된장의 경우 간장을 가르고 난 메주를 마쇄한 다음 여기에 소금, 간장 등을 혼합하여 발효시켜 만드는데 비해 공장식 된장은 메주 대신 순수한 미생물(*Aspergillus Oryzae*)이 배양된 코오지를 사용해 만든다. 코오지를 삶은 콩을 넣고 소금물을 넣어 용기에 틈이 생기지 않게 단단히 채우고 위에 소금을 뿌리고 비닐 또는 기름종이로 덮어 누름돌을 얹는다. 보통 25~30℃에서 2개월 정도 숙성하면 메주 중의 단백질 분해효소에 의해 아미노산이 생성되어 감칠맛이 난다. 내염성 유산균이 생산하는 유산, 내염성 효모가 생산하는 알코올이 상호 작용하여 숙성된다. 필요에 따라 숙성 중 뒤섞기를 하고 조합 후 마쇄, 살균, 포장과정을 거치면 개량식 된장 제품이 된다. 개량식장(醬)은 코오지 발명 이후 일본식장(醬)을 담그는 방법과 동일하게 되었다.

간장 제조

공장산 양조간장과 전통간장의 가장 큰 차이는 소맥을 이용한 코오지 제조과정이다. 증자한 탈지대두와 볶은 소맥을 동량으로 혼합하고 코오지을 접종하여 6개월간 발효숙성시켜 생간장을 얻는다. 이 생간장에 당류 등 식품첨가물을 첨가한 후 살균, 여과하여 만든 것이 양조간장이다. 산분해간장은 탈지대두나 소맥전분의 부산물인 글루텐을 염산으로 가수분해하여 아미노산으로 생성시키고 중화제로 중화시킨 후 액을 분리한다. 효소분해간장은 탈지대두와 소맥에 코오지를 제국하여 액을 간장덧을 만든 후 효소제를 첨가하여 만든다. 혼합간장은 양조간장, 산분해간장, 효소분해간장 등을 적당한 비율로 혼합하여 만든 간장이다.

간장의 종류

[농도에 따른 분류]

- **진간장** : 담근 햇수가 5년 이상 되어 맛이 달고 색이 진하여 약식, 전복초 등을 만드는데 쓰인다.
- **중간장** : 담근 햇수가 3~4년 정도된 장으로 찌개나 나물을 무치는데 쓰인다.
- **묽은 간장** : 담근 햇수가 1~2년 정도 되어 맑고 색이 연하여 국을 끓이는데 쓰인다.

[원료에 따른 분류]

- **조선간장** : 콩만을 원료로 하여 전분질을 사용치 않으며 주로 세균(*Bacillus subtilis*)에 의존해서 발효시킨다.
- **일본식간장** : 콩과 전분질을 혼합하며, 발효균도 곰팡이(*Aspergillus oryzae*)를 사용한다. 요즘 공장에서 대량 만들어지는 간장은 거의 모두 일본식의 간장이다.
- **어(魚)간장** : 어체나 그 내장을 원료로 하며, 특히 미생물의 힘을 빌리지 않고 자체의 효소에 의해서 분해 숙성된다. 어간장은 중국의 해안지방, 우리나라의 남해안지방, 일본, 동남아지방에 널리 분포하며 원료가 되는 어체는 그대로 이용하거나 머리와 내장을 제거하고 이용하기도 한다.

[제조법에 따른 분류]

- **재래식간장** : 순콩으로 간장을 만든다.
- **개량간장(양조간장)** : 콩밀로 제조하여 부산물인 된장이 나오지 않는다.
- **아미노산간장(화학간장)** : 산분해간장은 아미노산간장 또는 화학간장이라고도 하는데 우리나라 식품 위생법에서는 '산분해간장'이라 부른다. 즉 단백질 원료를 염산으로 가수분해 후 알칼리로 중화해서 짧은 시간에 간장을 제조한다.
- **무염간장** : 극도로 소금을 기피해야 하는 환자의 경우를 위해서 나온 것이 무염간장이다. 이것은 발효에 의해서 제조한다는 것은 불가능하며 산분해의 방법으로 제조할 수 있다.

고추장 제조

공장에서 주로 제조되는 방법으로 소맥분으로 증자하여 제국한 것에 증자된 밀쌀 또는 쌀, 찹쌀 등을 혼합하여 식염과 물을 가하고 수분을 약 50%로 조절하여 이를 마쇄하여 발효시킨 후 물엿, 고춧가루 등 첨가물을 넣어 60~70℃에서 살균하여 제품으로 낸다. 고춧가루를 숙성 전에 첨가하는 방법과 숙성 후에 첨가하는 방법이 있다.

청국장 제조

현재 공장에서 제조하는 청국장은 우량균주를 따로 배양, 접종하여 청국장을 제조한다. 청국장은 바실러스라는 균을 주로 하여 발효시키는데 이 균을 따로 위생적으로 배양하여 접종 사용한다. 즉 전통식 청국장은 볏짚에 묻어있는 종균을 사용하고 개량 청국장은 우량균주를 배양, 접종하여 발효시킨다. 우량균주를 사용하여 발효시킬 경우 품질의 균질화가 이루어져 대량생산이 가능하다는 장점이 있다.

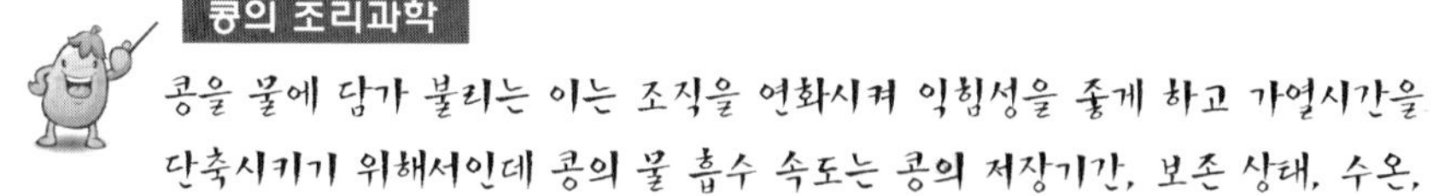

콩의 조리과학

콩을 물에 담가 불리는 이는 조직을 연화시켜 익힘성을 좋게 하고 가열시간을 단축시키기 위해서인데 콩의 물 흡수 속도는 콩의 저장기간, 보존 상태, 수온, 침지액의 종류와 양에 따라 다르지만 보통 19~24.5℃에서 5~7시간은 흡수 속도가 빠르고 20시간 정도에서 포화 상태에 이르게 되어 본래 콩 무게의 90% 이상 물을 흡수하게 된다. 콩을 가열하면 단백질 분해 효소가 작용하여 소화성이 높아지며, 생콩 특유의 비린내 성분이 휘발하게 된다. 가열 시에 팽윤 현상이 자엽보다 표피쪽이 빨리 일어나 주름이 생기는데 끓기 전에 냉수를 부어 온도를 조절하면 물이 잘 흡수되고, 설탕농도가 높은 경우에도 삼투압이 높아져 자엽이 수축하고 종피에 주름이 생겨 딱딱하게 되므로 설탕을 조금씩 나누어 넣으면서 서서히 설탕농도를 높이면 주름이 덜 생기게 된다.

3. 세계인의 식품, 콩

콩은 복잡한 가공과정을 거치지 않고 종자 그 자체로도 훌륭한 식품이지만, 실제로 서구에서 이용되는 경우에는 단백질과 지방을 분리하여 이용한 가공식품이 점점 많아지고 있다. 세계적으로 생산되는 대두박(soybean meal) 중 약 90%가 사료의 용도로 사용되고 있고, 식품가공에 이용되고 있는 양은 전체 콩단백의 0.5%, 대두유의 2.6% 수준이다.

한국, 중국, 일본, 인도네시아, 인도 등 아시아권에서는 콩식품이 일상식사의 일부분으로 단백질과 식물성 기름의 주된 급원이 되고, 콩을 이용한 여러 가지 발효성 장류는 동남아시아의 특징적인 만능 조미료이자 전통식품으로 중요한 위치를 차지하고 있다. 두유는 우리나라를 비롯하여 중국과 일본 등에서 콩국으로 마셔왔다. 북한에서는 '콩우유'라고 부른다. 두유는 콩의 수용성 추출액으로 외관이나 성분상 우유나 모유와 비슷하다. 전통적 두유는 콩을 수침 후 마쇄, 여과, 가열하여 비지(soy pulp)를 제거한 후 얻는다. 비지는 두유 제조 중 여과 후 생기는 부산물이다. 비지는 종피와 배아 등의 섬유질, 양질의 단백질, 지방, 무기질로 구성되어 있다. 비지는 그 자체로도 이용되나 최근 비지를 건조시킨 후 가루로 제조하여 발효식품, 조미식품, 제과제빵에 식이섬유 첨가제로 이용하고 있다.

세계의 콩 식품

▪ 중국 : 두반장, 쑤푸, 두시

중국에서는 대두와 그 가공품은 일상 식사 중에 자주 식탁에 오른다. 아침식사로는 콩국(불린 콩을 갈아서 만든 두유)을 먹는데, 아침 출근길에 노점이나 작은 식당에서 먹기도 한다. 점심과 저녁에

는 두부와 탕엽(湯葉두부피: 두유에 콩가루를 섞어 끓여 그 표면에 엉긴 엷은 껍질을 걷어 말린 식품), 두부간(豆腐干: 말린 두부(두부를 베로 싸서 향료를 넣고 찐 후 말린 것))이 요리에 자주 사용된다. 탕엽으로 고기나 채소를 김초밥처럼 말아서, 혹은 볶거나 쪄서 먹는다. 두부간은 얇게 썰어 볶은 음식에 자주 사용한다. 중국의 콩발효식품에는 두반장(豆飯醬), 쑤푸(sufu: 두부를 발효시킨 식품으로 조직과 풍미가 치즈와 비슷하여 중국치즈라고 함), 두시(豆豉)가 있다. 우리의 된장, 간장에 해당하는 함두시와 청국장에 해당하는 담두시가 있다. 매콤한 사천식 요리에 많이 쓰이는 두반장은 중국에서 제일 유명한 두부요리인, 마파두부에 곁들이는 소스로서 잘 알려져 있다.

일본 : 미소, 쇼유, 낫도

일본은 대표적인 장수 국가이다. 장수 비법은 발효된 음식을 즐겨 먹으며 또 강한 양념을 쓰지 않고 식재료 그대로의 맛을 내는 조리법을 이용한다. 그런 일본인의 식탁에서 빠지지 않는 식품이 바로 '낫도'이다. 우리나라 청국장과 비슷한 형태이긴 하지만 된장이 되기 전까지 발효시킨 음식으로 콩의 형태가 그대로 남아 있고 발효로 인해 점질물이 만들어진다. 뜨거운 밥에 낫도를 얹고 날달걀을 풀어 비벼먹는 것이 일반적인 방법이다. 낫도는 골다공증 예방, 변비, 다이어트에 효과가 있는 것으로 알려졌다. 최근에는 낫토를 지속적으로 섭취하면 고지혈증을 개선하는 효과가 있다는 연구결과가 발표되기도 했다.

미소[味噲]는 찐 백태를 쌀 또는 밀, 곡자(koji)를 소금과 함께 섞어서 만든 일본 된장이다. 누룩의 종류에 따라 쌀미소, 보리미소,

콩미소로 구분된다. 색깔에 따라 크게 적미소(赤味噌), 흰미소(白味噌)로 나뉜다. 우리나라를 비롯 동양 여러 나라의 공통된 콩발효식품이지만, 1904년 일본이 코지균을 상용화한 이래 현재 대다수의 나라에서 간장은 모두 일본식으로 생산하고 있다.

일본 식생활사에 있어서 주부식·분리시대인 야요이시대 부터 발효, 염장식품인 장이 쓰였다. 장에는 쌀, 보리, 콩 등을 발효시킨 곡장(穀醬)과 수조류를 소금에 절인 육장(肉醬), 그리고 열매, 채소, 해조 등을 소금에 절인 초장(草醬)의 3종류가 있다. 곡장은 후대의 미소와 소유(醬油), 육장은 후대의 시오가라(鹽辛)와 스시며, 초장은 후대의 쯔게모노의 원형이 된다. 콩을 이용한 음식으로는 두부를 가지고 만든 기누고시두부(錦豆腐), 고오리두부(氷豆腐), 덴가꾸(田藥) 등이 있다. 덴메이(天明) 2년에 만들어진 <두부백진豆腐白珍>(1782년)에는 두부로 만든 요리가 230여종이 소개되어 있다.

▪ 인도네시아 : 템페, 온쯤

인도네시아에서는 콩은 주로 두부, 템페, 식용유, 장류와 두유 등으로 이용되고 있다. 인도네시아 시장 어느 곳에서나 쉽게 볼 수 있는 두부는 가족기업이나 소기업으로 생산, 판매된다. 두부를 만들 때는 우리나라와 달리, 매우 작은 콩을 쓴다. 인도네시아의 대표적인 템페는 콩의 껍질을 버리고 물에 담가 불린 후, 라기(ragi)[1]를 섞어 하루나 이틀 정도 발효시켜 케이크 형태로 만든다. 라기 속에는 템페의 발효균인 라이조프스 곰팡이가 섞여 있다. 템페는 30℃의 온도에서 잘 발효가 되는데 하얗고 끈끈한 균사가 생기면 하얀

1) 효모(yeast)라는 뜻. 좁게는 인도네시아 전통 고체 이스트를 말하며, 넓게는 거의 모든 형태의 다양한 발효에 이용되는 발효 스타터를 의미한다

템페가 되고 저온에서 발효하면 회색이나 검정색의 균사가 생겨 검은색의 템페가 된다. 이들 균사는 콩에 서로 달라붙어 형태를 유지하게 한다. 템페는 생으로 먹지 않고, 간장을 발라 굽거나 얇게 썰어서 기름에 튀기든가 스프에 넣어서 먹는다.

브라질 : 페이조아다

콩은 항암 효과가 널리 알려져 있는 식품이다. 특히 검은콩 껍질에는 노란 콩에는 없는 글리시테인(glycitein)이라는 항암 물질이 들어 있다. 요즘 브라질에서 최고의 보양식으로 꼽히는 음식 중 하나가 검은콩으로 만든 '페이조아다(feijoada)'다. 이 음식은 원래 노예들의 일용식(日用食)이었으나 최근 콩에 대한 관심이 커지면서 새롭게 주목받고 있다.

아프리카 : 다와다와

서아프리카 일대의 사바나에서 만드는 전통발효조미료다. 원래는 로커스트 빈(메뚜기콩)이라고 불리는 아프리카산 검은콩을 원료로 만들지만 지금은 로커스트 빈이 모자라기 때문에 콩이 대체원료로 이용되기도 한다. 삶은 콩을 동글동글하게 빚어 햇볕에 말려서 만드는데 수프나 소스를 만들 때 넣어서 먹는다. 다와다와는 띄운 후에 단자 모양으로 뭉친 것을 손으로 눌러서 납작하게 만들고 햇볕에 건조시킨 보존성이 있는 식품이다. 로커스트 빈이 원래 검은색이기 때문에 제품도 검은빛을 띄고 있고 냄새는 극히 강렬하다. 다와다와는 소스나 스튜의 베이스로서 필수 불가결한 조미료이다. 다와다와의 발효균은 한국의 청국장균과 매우 비슷한데 "낙산취"의 강렬한 냄새가 있다.

네팔 : 키네마

청국장과 비슷한 음식으로 키네마가 있는데 네팔의 동부산악지대에 사는 기라토족들이 겨울에 즐겨 먹는다. 콩을 삶아서 으깬 다음 재를 섞어서 발효시킨 뒤 햇볕에 말리면 키네마가 완성된다. 냄새가 고약한 것이 특징이다.

태국 : 토-아나오

토-아나오란 썩은 콩이란 뜻이다. 삶은 콩을 바나나 혹은 산마의 잎으로 싸서 3~4일 동안 발효시켜 만든다. 잘 발효된 콩을 소금과 향신료를 넣어 찧은 다음 시루에 쪄서 양갱 모양으로 만들기도 하고 햇볕에 말려 쓰기도 한다. 냄새가 심한 게 특징이다. 태국 북부 산악지대에 사는 사람들이 즐겨 먹는다.

부탄 : 리비·잇빠

삶은 콩을 대광주리에 담고 천으로 덮어 습기 찬 방에서 실온으로 띄운다. 냄새가 나면 절구로 찧어서 단지에 넣고 다시 따뜻한 곳에서 숙성시킨다. 숙성시간은 1~3년이 걸리고 기간이 길수록 좋은 제품이 된다. 소금도 넣지 않는 상태에서 오랫동안 두게 되므로 냄새가 대단해서 이것을 만드는 소수민족들 사이에서는 싫어하는 사람들도 있어 극히 일부지방에서만 만들어진다. 리비·잇빠의 뜻은 "콩이 썩는다"이고 조미료로 이용된다.

인도 스자체

인도는 많은 인종과 종교가 섞여 있고 요리도 다양하다. 인도 음식의 단백질원은 여러 종류의 콩류와 우유, 버터, 요구르트 등

유제품이 대부분이다. 스자체는 인도의 앗사무 지방에서 만들어 먹는 음식으로 대바구니에 삶은 콩을 넣어 발효시킨다. 잘 발효된 콩은 절구에 찧은 다음 둥글게 뭉쳐서 바나나잎으로 싸서 말린다. 이렇게 말린 스자체는 오랫동안 보관하여 먹는다.[1)]

산업적 콩의 이용

콩단백질 분리 제품

콩의 불순물을 제거한 뒤 6~8조각으로 조분쇄하여 콩 껍질을 제거한다. 그 다음 자엽과 배아 부분을 얇게 박편화한 뒤, 용매로 지방을 추출한다. 이 때 지방 추출을 위한 콩의 수분 함량은 12% 이하인 것이 좋으며 만일 그 이상일 때에는 82℃ 이하에서 건조하여 수분 함량을 12% 이하로 조절한다.

- 콩가루(soy flour)

콩가루는 탈지한 대두박이나 탈지하지 않은 콩을 미세하게 분쇄하여 분말화한 것으로 글루텐이 함유되어 있지 않고 단백질 함량이 많다는 면에서 밀가루와 구분되며 섬유소가 함유되어 있다는 면에서 탈지분유와 다르다. 전지 콩가루는 콩 껍질을 제거한 콩을 마쇄한 것이므로 일반 성분은 콩의 자엽 부분의 성분 조성과 비슷하며 탈지 콩가루는 지방 성분을 제거하였기 때문에 지질 함량이 거의 없고 단백질 함량이 높다. 콩가루의 주요 품질 규정을 보면 탈지 콩가루는 지방 함량이 1% 이하여야 하며, 전지 콩가루는 최소한 18%의 지방을 함유하고 있다.

---

1) 조정순, 다른 나라의 콩 이용 음식, 콩, 한국콩박물관건립추진위원회편, 고려대학교 출판부, 529-578(2005)

- 농축 콩단백(soy protein concentrate)

농축 콩단백은 단백질과 불용성 탄수화물로 구성된 것으로 1959년 미국에서 최초로 상품화되었다. 농축 콩단백의 품질규격은 표피를 제거한 콩에서 대부분의 지방과 수용성 비단백태 물질을 제거하여 단백질함량이 건량 기준으로 70% 이상 되는 제품이다. 농축 콩단백 중 질소 용해도가 65% 이상인 것은 유화력과 결수력이 좋기 때문에 소시지와 같은 식품에 첨가하여 이용하고 있으며, 카제인이나 탈지우유를 필요로 하는 제품, 즉 육제품, 영양 강화음료, 스프베이스 등에 사용된다.

- 분리 콩단백(soy protein isolate)

분리 콩단백은 일찍이 1903년경 '소이 카제인(soy casein)'으로 불리워진 것으로 종이에 피막을 입힐 때 우유 카제인 대신 색소와 결합시키는데 처음으로 사용되었다. 분리 콩단백의 이용은 효소에 의하여 단백질을 변성시키거나 분리 콩단백을 그대로 과자, 빵, 소세지 등에 첨가함으로서 단백질 강화 원료로 이용되고 있다. 분리 콩단백은 단백질 함량이 건물량으로 90% 이상이 되도록 제조한 분말 콩제품을 말한다. 제품의 품질은 지방질이 전혀 없어야 하며 섬유질의 함량이 극히 적은 것이 특징이다.

콩 단백 제품의 이용

탈지 콩가루, 농축 콩단백, 분리 콩단백 등 콩단백 제품은 최근 소시지, 햄버거 고기 등 육가공 식품과 치즈, 커피 크림 등 우유 가공 식품, 그리고 빵류에 첨가하여 단백질 강화 및 원료비의 절감 목적으로 많이 사용되고 있다. 그러나 효과적인 콩단백질의 첨가를 위해서는 몇 가지 고려해야 할 사항이 있다. 즉 고기, 달걀, 우유와

같은 가격이 비교적 비싼 동물성 단백질 제품이나 식빵에 첨가할 때는 이들 식품의 품질 특성인 냄새, 맛, 텍스처(texture), 색 그리고 형태에 좋지 않은 영향을 주어서는 안 된다. 이들 제품에 사용하는 콩단백 제품은 냄새 면에서 분리 콩단백이 가장 유리하며 저장 기간이 짧은 것일수록 좋다.

육가공식품

오래전부터 서양에서는 육가공제품은 가격이 높은 고기 대신 비교적 저렴한 콜라겐이나 식물성 단백질을 일부 첨가하여 왔다. 햄버거 고기에는 농축콩단백이나 조직콩단백을 증량 원료로 첨가하며 소시지 제품에는 탈지 콩가루나 농축 콩단백을 첨가한다. 이렇게 육제품에 콩단백질을 첨가하면 첫째 육류 제품의 가격을 저렴하게 하고, 조리할 때 수축됨을 감소시키며, 육류 식품의 유화력과 안정성을 향상시킨다. 또 고기 입자들의 결착 능력을 향상시키고, 수분 흡수 능력을 높여 주어 씹을 때의 촉감, 견고성 등의 텍스처를 향상시키는 효과가 있다. 햄버거 고기에는 조직 콩단백 또는 농축 콩단백을 25~30%까지 첨가할 수 있어, 원료 가격을 20~30% 정도 절감시켜 준다.

콩고기 만드는 방법

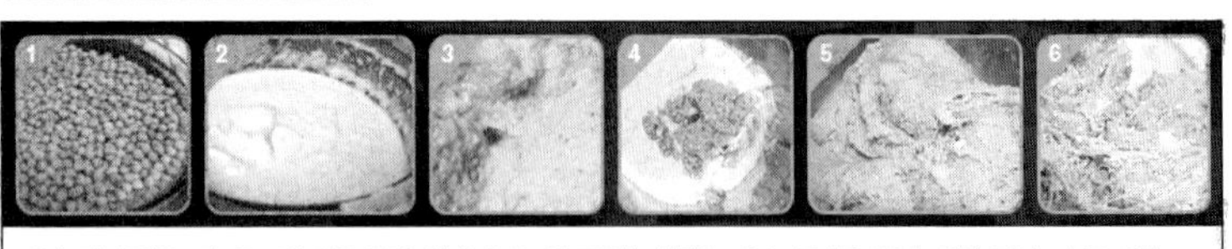

[재 료] 불린콩 1컵, 물 3컵, 땅콩 반컵, 참깨, 호두, 잣 조금씩, 생강즙 1큰술, 들기름 3큰술, 간장 2큰술, 소금 1큰술, 우리밀가루 반컵, 글루텐 1컵

❶ **콩 갈기**: 콩을 하룻밤 불려 살짝 삶은 후 갈아낸다.
❷ **견과류 갈기**: 땅콩 등 견과류는 분쇄기로 갈아낸다.
❸ **견과류섞고 간하기**: 콩물에 견과류를 넣고 간장, 소금, 생강즙, 들기름 등으로 간을 한다.
❹ **반죽하기**: 밀가루를 섞고 글루텐을 조금씩 넣어가며 반죽한다. 농도가 너무 묽거나 되지 않도록 조절을 하며 반죽한다.

우유가공식품

콩 단백질의 첨가가 가능한 유가공식품은 아이스크림, 치즈, 커피크림, 휘핑크림 등이 있고 두유와 우유를 혼합한 혼합음료도 있다. 이 혼합음료는 두유에 풍부한 리놀레산, 리놀레산, 리신, 페닐알라닌을 우유에 강화시킴으로써 난황보다 우수한 영양성분 조성을 갖게 할 뿐만 아니라 락토오스의 함량을 낮게 하는 효과도 있다. 이 외에 분리콩단백을 설탕 또는 옥수수시럽, 무기물, 지방, 비타민, 그리고 유화제 등을 물에 녹여 균질화한 음료도 있다.

빵 및 스낵류 식품

콩 단백 제품들을 빵이나 스낵류에 첨가하면 단백질 영양가가 향상될 뿐만 아니라 이들 제품의 보수력이나 흡수력을 향상시키는 장점이 있다. 빵에 12%의 탈지콩가루를 첨가하게 되면 빵의 단백질 양이 약 50% 증가되는 효과가 있을 뿐만 아니라 단백질 효율치도 높아지게 된다. 빵과 스낵류에 콩단백질 제품을 첨가하면 이들 제품의 보수력이 향상되고 빵이 딱딱해지는 현상이 감소하며 반죽이 용이해지고, 케이크의 부드러운 성질이 향상된다. 또한 빵 표면의 갈색화가 빠르게 되며 도넛의 기름 흡수력을 향상시키고 전반적으로 영양적 가치를 높여주는 효과가 있다.

콩 단백질의 조직화

콩 단백질 제품인 콩가루나 탈지 콩가루, 농축 콩단백, 분리 콩단백은 그대로 조리하여 섭취하기에는 식감이 떨어지기 때문에 조직화하는 기술이 개발되었다. 콩단백질을 조직화시키는 방법은 압출, 방적(spinning), 스팀조직화(steam texturizing) 등이 있다. 조직을

갖게 되는 것은 압출기에서 또는 증기에 의한 가열처리에서 콩 단백질의 수소결합이 파괴되어 단백질이 펴지기 때문이다. 펴진 단백질은 고압 하에서 긴 형태로 재차 정렬하게 되며 이를 냉각하면 새로운 수소결합이 형성되어 단백질 분자의 재배치가 일어나고 분자 상호간에 교차결합하게 되어 새로운 조직 구조를 안정화시켜 준다. 그렇게 조직이 부여된 콩 단백제품은 섭취하기 좋게 되고 기호성을 향상시켜, 씹힘성과 맛이 육류와 근사하게 되는 것이다.

콩기름의 이용

콩이 유럽과 미국에 도입된 초기에는 압착에 의한 방법으로 콩기름을 생산하였다. 그러나 점차 사회가 발달하고 인구가 증가함에 따라 기름의 수요와 소비가 증가하면서 대량생산이 가능한 원료에 대한 필요성이 제기되었다. 이러한 시점에 유기 용매를 이용한 추출 방법(solvent extraction)이 개발되어 식물의 종자에서 저렴한 가격에 정제 유지를 대량생산할 수 있게 되었다. 대두유는 생산 및 공급이 가장 많은 유지로 쇼트닝, 마가린의 원료로, 식용유로서 굽거나 튀김용, 그리고 요리용 등으로 다양하게 이용된다. 콩에는 20% 정도의 기름을 함유하고 있으며 올레인산, 리놀레산, 리놀렌산 등 불포화지방산을 85% 정도 갖고 있어 폭넓은 온도에서 액체상으로 존재하여 응용범위가 넓다. 천연항산화제인 토코페롤이 많아 산화 안정성 및 풍미에 기여한다. 유지는 온도에 따라 고체에서 액체로 또는 그 역으로 전환이 가능하기 때문에 제과, 제빵에 응용되며 특히 튀김 쇼트닝으로서 활용을 기대할 수 있다. 세계적으로 콩기름의 생산은 콩의 생산과 같이 계속적으로 증가하는 추세이다.

콩기름의 용도

유지를 일정한 온도로 가열한 후에 적당한 촉매(보통 니켈)를 넣고 교반하면서 수소가스를 흘려주면 일어나게 된다. 온도, 압력, 촉매의 종류, 수소의 양 등의 조건이 충족되면 수소첨가 반응은 신속하게 일어나게 된다. 이러한 반응이 계속 진행되면 최종적으로는 트리글리세라이드의 모든 이중결합이 포화되어 완전한 수소첨가가 되는 것이다. 이러한 경우에는 비교적 융점이 높은 고체유지를 얻을 수 있으며(경화유), 식품가공에 이용된다. 경화유의 대표적인 마가린 및 쇼트닝은 가소성 유지, 산화 안정성 증대에 이용되거나 버터 등의 대용 유지로 활용된다.[1)]

콩기름 부산물의 제조와 용도

레시틴

콩기름은 초기에 대두를 뜨거운 핵산으로 추출한다. 그 결과 미세물, 단백질, 금속불순물 등을 제거하기 위하여 여과를 거쳐야 한다. 핵산은 증류하여 제거하고 조유는 수세처리하면 인질을 함유한 유지불용분이 생기게 된다. 레시틴의 산업적 용도의 특성은 유화력과 보습력이 있다. 동물사료용, 제빵제품, 캔디, 초콜릿, 아이스크림, 마카로니, 라면, 마가린, 식용유지, 잉크 염료, 제약 등에 많이 쓰인다.

토코페롤

토코페롤은 1922년 에반스(Herbert M. Evans)에 의해 분리되었다. 토코페롤의 원어는 tocos(자손)와 pherein(낳다) 및 ol(알코올)의

1) 이경일, 콩기름과 그 부산물, 콩, 한국콩박물관건립추진위원회편, 고려대학교출판부, 691-748(2005)

합성어로서 출산과 관계가 있는 매우 중요한 기능성분으로 알려져 있다. 토코페롤은 산패를 방지하는 항산화제의 역할과 필수 영양소인 비타민 E의 생리활성기능을 갖고 있다. 토코페롤의 추출 양은 원료 대두의 생태학적 조건이나, 탈취공정조건, 즉 탈취온도, 탈취시간에 따라 다소 변화가 있다.

대두박의 이용

사료로 이용되는 것은 대부분이 대두박의 형태이며 그 외에 소량의 대두유가 사료의 에너지 급원으로 사용되고 있다. 대두박이 사료의 원료로 사용된 역사는 콩기름 착유의 역사와 같다. 초기에는 압착이나 추출방법에 의해 착유하고 남은 부산물인 콩깻묵이 그대로 가축에게 제공되었다. 하지만 1930년대 이후 대두박을 토스팅(toasting)하여 항 영양인자를 제거하는 방법이 제시된 이후에는 더 높은 영양가의 대두박이 제조되었다. 대두의 착유과정, 대두유 정제과정, 대두단백질의 정제과정에서 대두박(soybean oil meal), 대두피(soybean hulls), 대두레시틴 등이 부산물로 생산된다.

우리나라의 식품 외 콩의 용도

- 콩깻묵 : 콩에서 콩기름을 추출하고 남은 콩깻묵은 가축의 사료를 만드는 중요한 원료로 사용되었다. 가축을 사육하는데 필요한 사료의 단백질급원으로 과거에는 동물성 급원을 주로 사용하였지만 육류에 대한 수요가 증가하면서 동물성 급원의 단백질만으로는 필요한 수요를 충족할 수 없게 되었다. 따라서 질이 우수하고 안정적인 공급이 가능하고 가격이 저렴한 콩깻묵을 이용하였다. 냇가에서 고기 잡을 때 콩깻묵을 이용했다.
- 비료 : 아주 황폐한 땅에는 콩을 조금 자라게 두었다가 뿌리째 갈아엎어 거름이 되게 한다. 두번째 김맬 때 논 1마지기마다 콩 4~5말을 뿌려주되, 적게는 1~2말이라도 흩어 뿌려주면 삼복더위에 진흙이 끓어오르면서 콩이 다 녹아서 비료가 된다.

- 세정제 : 두부를 만들고 난 순물로 세수나 목욕을 하면 피부가 부드럽다. 동상에 걸린 손발을 담그기도 한다.
- 도장밥 : 두유를 이용하여 붉은 염료를 갈면 오래되어도 새롭다고 하였다.
- 우의 : 처음에는 진유(眞油: 참기름)를 발라 그늘에 말리고, 계란, 활석가루, 송진가루, 법유(法油: 들기름), 두즙을 넣어 곱게 갈아서 비옷 안팎에 4, 5차에 걸쳐서 바른다.
- 콩댐 : 흰콩 1과 물 6의 비율로 불린 다음 곱게 갈아서 베보자기로 꼭 짠다. 콩물4에 들기름 1을 섞어서 준비, 마루에 콩댐을 할 땐 콩물 2와 들기름 1로 하여 3번 이상 칠해준다.
- 호롱불 : 한지, 솜, 마사, 삼 등을 꼬아 심지를 만들고 콩기름, 아주까리기름 등을 연료로 썼다. 심지가 두 개인 쌍심지는 매우 밝은 반면, 기름을 많이 먹는다.

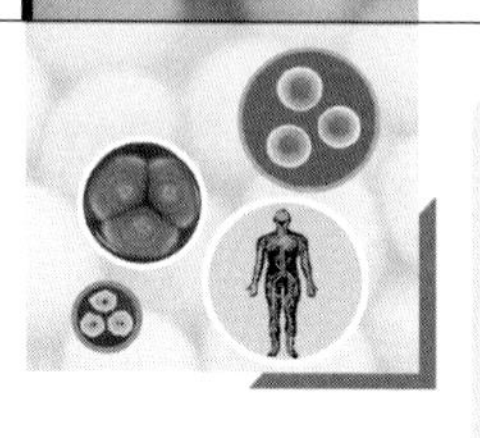

콩의 영양과 기능성

1. 콩의 구조와 구성
2. 콩의 영양과 생리활성물질
3. 콩 식품의 기능성
4. 질환별 콩의 효능

1. 콩의 구조와 구성

현대인들의 식품선택과 식사패턴에 대해 전 세계적으로 경종을 울린 사건이 1970년대에 미국에서 있었다. 미국인들의 식생활에 대해 조사한, 미 상원의 영양문제 특별위원회에서 2년간에 걸친, 5천여 페이지에 달하는 방대한 보고서를 발표했을 때였다. 처음에 영양문제위원회는 기아문제에 관심을 가졌지만 점차 식품과 영양, 즉 식생활이 건강에 미치는 영향에 대해 조사하게 되었다. 1977년 1월 맥거번 위원장은 "분명한 사실은 우리들의 식생활 양상이 지난 반세기 동안 부정적으로 변해왔으며, 그 결과 우리들의 건강에 지대한 나쁜 영향이 있다. 지방이나 설탕 그리고 소금의 지나친 섭취는 여러 가지 치명적인 병들 가운데서도 특히 심장병, 암, 뇌졸중과 직접적인 연관성을 가지고 있다. 미국인의 10대 치명적인 질병 가운데 6가지는 그 원인이 우리들의 식생활과 연관되어 있다."고 발표하였다. 이 위원회의 보고 이후 미국인들의 식생활은 물론 전 세계적으로 건강에 대한 관심과 웰빙열풍이 확산되는 계기가 되었다. 미국에서 콩이 식용으로 연구되고 제품화가 활발하게 시도된 것도 이때부터였다.

콩의 구조와 자엽의 구조

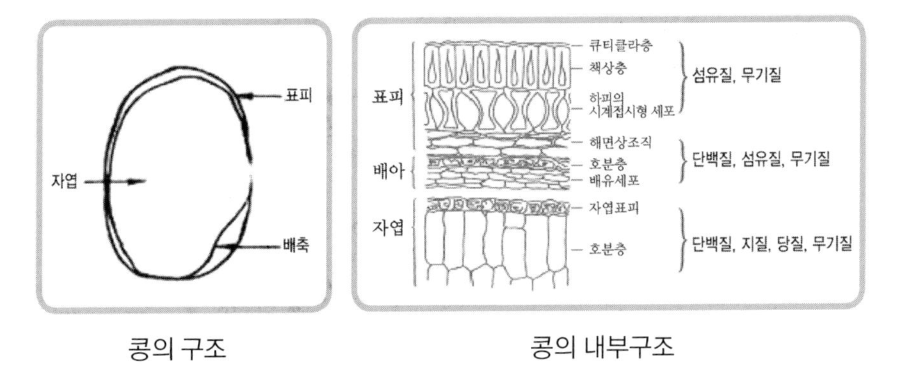

콩의 구조　　　　　콩의 내부구조

콩의 구조

콩은 결실시기가 되면 콩깍지를 형성하면서 그 안에 종자를 만든다. 콩의 종자는 껍질과 자엽과 배축(씨눈)으로 구성되어 있다. 콩의 일반 조성을 부위별로 알아보면 껍질 무게는 전체 콩 무게의 약 8.3%, 배축은 2.1%이고 나머지 약 90%는 자엽 부분이다. 이러한 구성 비율은 콩의 크기와 품종에 따라 달라진다. 콩의 껍질 성분은 대부분(86%)이 섬유질과 가용성 질소물이고 식용의 대상인 자엽 부분은 40% 이상의 단백질(건량)과 20% 정도의 지질이 함유되어 있다. 자엽은 떡잎이 되는 부분이다. 떡잎이 갈라지는 곳에는 생장점이 있어 그곳에서 싹이 자라서 콩 줄기와 잎들이 나오게 된다. 배축은 뿌리줄기가 되고 더 자라면 뿌리가 된다.

2. 콩의 영양과 생리활성물질

동물성 식품의 섭취가 높은 서양인에서 만성퇴행성 질환이 주요 사망원인이 되고 있는 반면, 동양인들은 이러한 질병 이환율이

서양인에 비해 낮은 것으로 나타나고 있다. 이러한 차이는 동양인의 식사에 많이 포함되는 콩의 섭취여부와도 관계가 있을 것으로 보고 있다. 콩 식품은 다른 식물성 식품에 비하여 지질 함량도 높고 콩 식품을 고기나 유제품의 대체 식품으로 활용할 경우 포화지방산과 콜레스테롤 함량도 낮아지므로 우수한 건강식품이다.

콩에는 3대 영양물질인 단백질, 지방, 탄수화물이 각각 40%, 20%, 18% 정도 들어 있으며 이외 비타민, 미네랄, 섬유소를 함유하고 있다 최근에는 이소플라본, 레시틴 등 콩의 생리활성물질까지 깊이 연구되고 있다. 우리 몸은 수분을 제외하면 단백질이 주 성분으로 되어 있다. 콩이 가진, 식품으로서의 가장 큰 가치는 바로 단백질 공급원으로서의 역할이다. 그래서 서구에서는 일찍부터 '밭의 고기'라고 불려졌던 것이다. 콩은 고기보다 단백질 함량이 2.5배나 많다. 콩 단백질이 분해되면 펩타이드가 만들어지고 이것이 더 분해되면 아미노산이 된다. 콩 단백질뿐만 아니라 콩 펩타이드의 효능도 연구대상이다. 콩 단백질은 혈중 콜레스테롤 수치를 낮춰 동맥경화, 심장병 등 혈관 질환의 발생 위험을 낮춰준다.

콩 얼마나 좋은가

예전에는 단백질의 영양가 판정방법으로 단백가(protein score)를 이용하여 동물성 단백질은 우수한 단백질이고 식물성 단백질은 질이 낮다고 평가되었다. 하지만 최근에는 단백질의 질적 평가를 아미노산가(PDCAAS: Protein Digestibility Corrected Amino Acid Score)라는 새로운 방법을 이용하여 단백질 속에 들어 있는 필수 아미노산의 양과 비율, 소화율 등의 효율성에 따라 단백질의 가치를 구하는 방법이다. 이 방법으로 계산하면 콩은 다른 우수한 단백질 식품과 비슷한 수준임을 알 수 있다(103p 그림 참조).

콩지질에는 몸에 좋은 불포화지방산이 절반이 넘는다. 총 지질 중 57%가 리놀레산이고 23%가 올레산, 그리고 8%가 리놀렌산으로 되어 있다. 리놀레산과 리놀렌산은 식품으로 꼭 섭취해야 하는 몸에 좋은 필수지방산이며, 특히 리놀레산은 암세포 성장 억제 효과가 밝혀지고 있다. 암세포의 성장을 직접 억제하기도 하지만 암세포에 영향을 주어 암세포가 스스로 자살하도록 유도한다.

올리고당은 탄수화물의 일종으로 단당이 2~10개 정도 결합된 당질로서 콩올리고당에는 스타키오즈, 라피노즈 등이 있다. 콩올리고당은 장내에서 장내 미생물에 의해 2차 대사물인 아세트산, 피루브산 등 저급 지방산으로 분해되어 인체 유익균으로 알려진 비피더스(bifidus)균 및 젖산균의 증식 및 유해균의 성장억제에 관여하여 정장효과를 나타낸다.[1)]

콩, 우유, 쇠고기, 백미의 필수아미노산 조성(mg/100g 가식부위)과 단백가[2)]

필수아미노산	아미노산표준구성	콩	백미	쇠고기	우유
아이소루이신	270	336	322	327	407
루이신	306	482	535	512	626
라이신	270	395	236	546	496
페닐알라신	180	309	307	247	309
메티오닌	144	84	142	155	156
트레오닌	180	246	241	276	294
트립토판	90	86	65	73	90
발린	270	328	415	347	438
단백가	100	73	72	83	78

1) 승정자, 콩 음식의 영양가와 기능성, 콩, 한국콩박물관건립추진위원회편, 고려대학교출판부, 579-660(2005)
2) 김우정, 콩의 가공특성, 콩, 한국콩박물관건립추진위원회편, 고려대학교출판부, 217-312(2005)

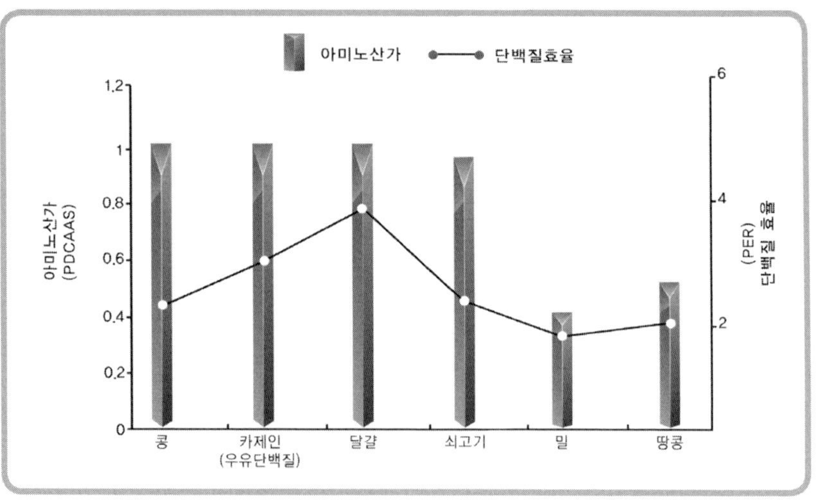

PDCAAS(소화율이 고려된 아미노산)와 단백질 효율 비교

삶은 콩 1/2컵에는 1일 철분 권장량의 44%가 들어있으며 다량의 칼슘과 마그네슘, 아연도 들어 있다. 또 비타민 B_1,비타민 B_3,비타민 B_6도 상당량 들어 있기 때문에 훌륭한 복합B비타민원이 된다. 콩에는 식이섬유질이 상당량 들어 있는데 동물과 달리 인간에게는 섬유질을 소화시킬 효소가 존재하지 않는다. 그러나 섬유질 중 불용해성 섬유질은 장내 연동작용을 촉진하고 중금속 등 유해물질을 흡착하여 배설시키는 역할을 한다. 그래서 변비 또는 대장암을 예방하게 된다. 반면 수용성 섬유질은 급격한 혈당 상승을 억제하여 당뇨병 예방에 도움을 준다.

콩의 생리활성물질

이소플라본(isoflavon)

이소플라본은 식물이 외부 환경에 노출되면 스스로를 방어하는

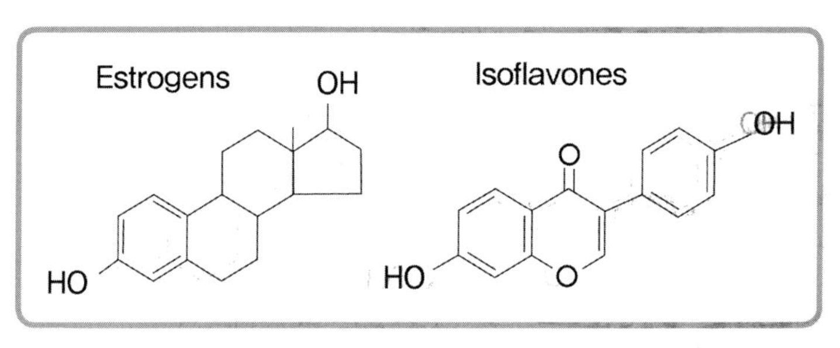

이소플라본과 에스트라디올(estradiol)의 구조[1)]

물질이다. 콩에는 어떤 식품보다 이소플라본이 많이 들어 있는데 어렵지 않게 콩을 먹을 수 있다는 것도 장점이다. 이소플라본은 '피토 에스트로겐(phyto-estrogen)'이라고도 하는데 '피토(phyto)'는 식물이란 뜻이고 '여성 호르몬(estrogen)'과 비슷한 역할을 한다고 해서 이름이 붙여졌다. 이소플라본은 인체 내에서 상황에 따라 에스트로겐 효과를 조절하는 역할을 한다. 즉 여성 호르몬의 분비가 적으면 콩의 이소플라본이 그 역할을 대체하고 여성 호르몬이 지나치게 많이 분비되면 그 양을 조절한다. 이소플라본은 골밀도를 높여 뼈를 튼튼하게 해주는 역할도 한다. 이소플라본의 하루 권장량은 아직 정해지지 않았지만 매일 50~100mg을 섭취하는 것을 권장하며 두부 1모에 150mg, 두유 1팩(200mL)에는 30mg, 된장(15g)에는 5.5mg 가량의 이소플라본이 들어 있다.

레시틴(lecithin)

인과 지질이 결합되어 있어 인지질이라고도 한다. 레시틴은 인체

1) Setchell KDR, Phytoestrogens: the biochemistry, physiology, and implication for human health of soy isoflavons, American Journal Clinical Nutrition, 68, 1333S~1346S(1998)

에서 세포막을 구성하고 콜레스테롤을 유화시켜 주며 콜레스테롤의 생리활성 물질인 호르몬을 도와주는 역할을 한다. 또한 신경전달물질이 되는 아세틸콜린의 원료가 되어 기억력을 증진시키는 중요한 역할을 한다.

사포닌(saponin)

콩을 물에 끓이면 하얀 거품이 뜨는데 이 거품 성분이 바로 사포닌이다. 사포닌은 위장에서의 콜레스테롤의 흡수를 방해함으로써 콜레스테롤을 조절하는 역할도 한다. 또 사포닌은 노화를 지연시키는 기능성물질로도 알려져 있다. 사포닌의 특성은 물과 기름에 모두 잘 녹는 것으로 이 같은 성질은 지질산화에 의해 만들어지는 과산화지질의 형성을 억제해 암, 동맥경화 등의 질병 발생률을 낮추어 준다.

폴리페놀(polyphenol)

항산화 및 항암특성을 보이는 폴리페놀 성분은 식물성 화학물질(ptyto-chemical)의 하나로 콩을 비롯해 여러 종류의 식물에 들어 있다. 폴리페놀은 인체 내에 들어오는 화학 쓰레기를 청소하는 기능이 있다. 즉 종양을 일으키는 화학성분을 방해해 인체 내 암세포의 성장을 억제하는 중요한 역할을 한다.

피트산(phytic acid)

피트산은 유해중금속과 함께 철의 흡수를 방해하는 작용이 있다고 보고되었지만 오히려 철분과 결합, 체내 철의 함량을 일정하게 유지시켜 줌으로써 활성산소의 형성을 억제시키는 효과가 있는 것으로 최근 보고되고 있다. 특히 결장암을 예방하는 효과가 우수하다.

피니톨(pinitol)

피니톨은 탄수화물의 유도체로서 체내에서 인슐린 유사작용을 하는 물질로 알려져 있다. 피니톨을 단기간 또는 장기간 섭취 시 당뇨 환자에게서의 섭취 효과를 조사한 결과 부작용 없이 식후 혈당의 급격한 상승을 억제하여 당뇨병의 예방 및 치료에 이용이 가능한 성분으로 연구되고 있다.

3. 콩 식품의 기능성

두 유

두유에는 콩에 함유된 성분이 거의 모두 들어 있다. 콩국이라 하여 옛부터 우리민족은 여름에 음료나 국수를 말아먹는데 애용해 왔으며 오늘날에는 우유 대용 식품으로 이용하고 있다. 두유는 생체막을 구성하는 다가불포화지방산의 주요 급원이며 지용성 비타민의 주요 급원식품이다. 특히 항산화 영양소인 비타민 E의 함량이 100g당 93.7 mg 함유되어 세포노화를 억제하는 기능이 있다. 두유에는 콩 단백질과 같은 영양성분뿐만 아니라 식이섬유소, 올리고당, 사포닌, 이소플라본의 생리활성물질이 함유되어 있다. 또한 유당(lactose)이 들어있지 않아 유당불내증[1]이나 우유알레르기가 있는 사람에게 부작용없이 충분한 영양을 공급할 수 있어 우유 대체 음료로서도 가치도 크다.

두 부

콩의 성분중 지질과 단백질이 두부 및 두유에 가장 잘 옮겨진다.

1) 유당불내증 : 우유에 들어있는 유당(lactose)을 분해하지 못해 소화불량을 일으키는 증세. 한국인의 70% 정도가 유당불내증을 가지고 있다.

따라서 콩 성분의 이용면에서 보면 단백질과 지질함량이 높은 것이 두부원료로 좋다. 두부의 일반 성분은 수분 80~85%, 단백질 약8.5%, 지질 약 5.5%, 당질 1.5%로서 두부에는 체내의 신진대사와 성장발육에 절대적으로 필요한 필수아미노산 및 필수지방산이 함유되어 있으며, 칼슘, 철분 등의 무기질이 다량 함유되어 있다.

콩나물

콩이 발아되어 생장하는 과정에서 체내대사를 통하여 영양성분이 크게 달라진다. 종실에 들어있는 지질은 생장 과정 중 크게 감소하는 반면, 식이 섬유소는 증가하고 특히 비타민이 많이 증가한다. 예를 들어 비타민 C의 경우 콩에는 전혀 함유되어 있지 않으나 콩나물 생장 중 5mg/100g 가식부위 정도 합성되므로 비타민 C의 함유량이 크게 증가된다. 이외에도 콩나물로 생장하는 과정에서 비타민 K와 나이아신이 만들어진다. 지질이 산화하여 생성되는 유해한 과산화지질을 감소시켜 동맥경화를 예방하는 기능을 가진 비타민 B_2도 콩나물로 생장하는 중 2배로 증가하며, 비만과 변비예방에도 효과적인 식이섬유소의 함유량도 증가된다. 또한 다른 두류의 나물보다 단백질과 철을 많이 함유하고 있다.

콩나물에 들어있는 아미노산은 필수 아미노산(리신, 트립토판, 메티오닌 등)과 유리 아미노산인데, 이를 섭취함으로 피로해진 간의 기능회복에 긍정적인 효과를 줄 수 있다. 콩나물 잔뿌리에 많은 아스파라긴산은 알코올분해효소의 기능을 도와 숙취해소에 좋으며, 콩나물에 많은 칼륨은 나트륨의 작용과는 반대로 혈압을 낮추는데 도움을 준다.

콩기름(대두유)

다가불포화지방산을 다량 함유하고 있는 콩기름은 항산화영양소 중에서 비타민 E(토코페롤)의 주요 급원이다. 콩기름에는 다른 식물성유보다 필수지방산 및 비타민 E 등이 많이 함유되어 있고 균형있는 지방산 조성을 보이고 있어 영양성과 기능성을 고루 갖춘 식품이라 할 수 있다. 100g당 필수지방산이 55% 이상 함유되어 있으며 고도불포화지방산인 리놀레산(linoleic acid)과 리놀렌산(linolenic acid)이 이에 해당한다. 리놀레산(오메가6계) 함량이 52.6%이고 리놀렌산(오메가3계)은 7.94%에 달한다. 특히 콩기름은 오메가6계와 오메가3계의 조성 비율이 8.21:1로서 세계 각국에서 권장하고 있는 바람직한 비율인 4~10:1에 적합하여 영양적으로 균형을 이루고 있다. 반면, 콩기름에 함유된 고도불포화지방산은 총콜레스테롤과 LDL 콜레스테롤의 감소 효과가 여러 연구에서 밝혀져 건강면에서 좋은 유지라 할 수 있다.

된 장

된장의 발효기간이 길어질수록 유리지방산, 이소플라본, 리놀레산 등의 함량이 증가되어 생리활성 효과가 커지며 발효과정 중에 형성된 갈변물질(갈변물질은 아미노산, 펩타이드, 단백질 등의 아미노기와 포도당 같은 환원당의 마이얄반응(Maillard reaction)에 의해 멜라노이딘(melanoidin)이라는 색소물질이 생겨나는데서 비롯된다), 여러 종류의 인지질, 토코페롤, 아미노산, 펩타이드 등은 항산화 기능에 기여한다. 된장의 바실러스속 균주는 강력한 혈전용해효소(subtilisin)를 분비하는데 전통된장의 혈전용해능은 상업용 된장보다 10배 높으며, 청국장이나 낫도보다는 3~4배 높은 혈전용

해능을 보인다. 바실러스균은 유기산도 생성하여 정장작용 효과도 보여주며 바실러스균의 포자는 살아서 장까지 갈 수 있으므로 그곳에서 발아하여 장에서 유익균(probiotics)으로서의 활성을 갖는다. 된장에 함유된 리놀레산은 암세포 성장 억제효과 뿐 아니라 암세포에 영향을 주어 암세포가 스스로 자살하도록 유도하여 항암작용을 나타낸다. 콩 발효 식품은 콩에 존재하는 프로바이오틱(유익균)인 바실러스균 등에 의해 암예방, 항산화 등 기능성이 우수하다.[1)]

청국장

청국장은 순수하게 세균의 힘만 빌려 콩 단백질을 분해한다. 콩 발효식품 중 가장 짧은 기간에 완성되며, 우리나라 청국장은 고초균(*Bacillus subtilis*)이 콩을 발효시켜 만든 것으로 100g당 1천 억마리 이상의 소화효소균이 생겨 소화흡수율이 향상될 뿐만 아니라 발효과정을 거치며 본래 콩에 함유되었던 좋은 성분에 기존 콩에 없던 유익한 생리활성물질이 더해져 영양가치가 높아진 우수한 전통식품이 된다. 청국장 발효의 주역인 바실러스균주는 장내 부패균의 활동을 약화시키고 병원균에 대한 항균작용도 나타낸다. 끈적끈적한 점질물은 바실러스 서브틸리스로부터 생산된 폴리글루탐산(polyglutamic acid)이라는 물질인데 화학적으로 감칠맛을 내는 아미노산인 글루탐산이 약 5,000개 연결된 펩타이드와 과당이 결합한 당단백질의 일종이다. 칼슘의 흡수를 촉진하여 골다공증 예방과 치료에 효과가 있을 뿐 아니라 정장작용 및 장내가스 발생을 감소시킨다. 청국장의 생리활성 물질로 이소플라본, 피트산, 사포닌, 트립

1) 박건영, 콩 발효식품의 건강기능성, 콩, 한국콩박물관건립추진위원회편, 고려대학교출판부, 407-454(2005)

신 저해제, 토코페롤, 불포화 지방산, 식이섬유소, 올리고당, 항산화물질, 혈전용해효소, 비타민 K 등이 있다. 효소를 섭취하려면 끓여 먹는 것보다는 생청국장이 좋다. 하지만 청국장에는 효소 이외에도 다른 건강기능 물질들이 많아 찌개로도 좋은 식품이다.[1)]

고추장

고추장에는 콩(메주)의 함량이 10% 내외 들어 있다. 고추장은 가장 늦게 전통장류에 편입되었지만, 가장 먼저 세계화되고 있는 장류가 되었다. 고추장은 매운맛 성분인 캡사이신, 색소성분 캡산틴(카로티노이드) 등이 콩 성분에 추가된 기능식품으로써 메주로부터 암 예방효과와 고춧가루의 캡사이신으로부터 체중감소와 지구력 증진효과, 캡산틴으로 부터는 항산화효과를 기대할 수 있다. 흡수된 캡사이신이 혈액을 통하여 중추신경을 자극하고 교감신경을 통해 부신피질로부터 아드레날린 등의 분비를 촉진하여 지방을 연소시키기 때문에 캡사이신은 운동했을 때와 거의 같은 정도로 에너지 대사를 촉진하여 비만을 예방할 수 있다.

발효로 인해 증대되는 효과

콩과 쌀 혹은 밀가루를 사용하여 만드는 간장이나 된장은 발효하는 과정에서 메주에서 자란 곰팡이가 만든 효소에 의해서 콩 속에 있는 단백질이 분해되어 구수하고 감칠맛을 내는 여러 아미노산이 만들어지게 된다. 함께 넣은 쌀이나 밀가루의 전분도 분해되어 단맛을 내는 포도당이나 맥아당으로 된다. 이들 아미노산과 당이 어울려 독특한 맛을 창조하며 고추장은 여기에 매운 맛이 추가되어

1) 김한복, 청국장 다이어트&건강법, 휴맨앤북(2003)

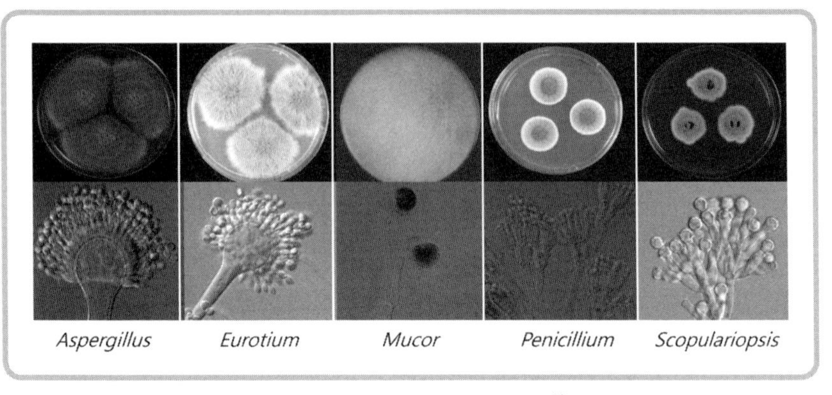

콩 발효에 관여하는 미생물들[1)]

오묘한 향신조미료가 된다. 발효식품은 이와 같이 미생물들이 원료의 성분을 분해시키기도 하나 분해된 산물이 다른 성분과 결합하여 성질이 다른 또다른 물질을 만든다.

소화흡수율의 증가

콩의 소화흡수율은 65%, 두부와 된장은 85%, 두유와 청국장의 소화흡수율은 95%에 달한다. 콩단백질의 영양적 품질은 가열 또는 발효가공에 따라 크게 변한다. 가열시킨 콩 제품이나 발효식품의 단백질 품질이 현저히 상승되는 것은 단백질 분해효소의 기능을 억제하는 트립신 인히비터(trypsin inhitor)가 열처리나 발효 중 불활성되거나 분해되었기 때문이다. 콩에 함유된 이소플라본(isoflavone)은 항암·골다공증 예방과 같은 기능을 수행하는 유용 생리활성물질로 알려져 있다. 이소플라본은 식품 중에 이소플라본과 당류가 결합한 배당체 또는 당류가 결합하지 않은 아글리콘(aglycone) 형태로

1) 홍승범, 국립농업과학원(2011)

존재하게 되는데 아글리콘은 배당체보다 생체 이용성이 우수한 형태로 알려져 있다. 두부, 콩나물에서는 콩과 마찬가지로 대부분 배당체 형태로 존재하지만, 된장과 같은 발효식품에는 많은 양의 이소플라본이 분해되어 생체 이용성이 우수한 식품이 된다.

비타민의 증가

원료 콩에는 비타민 B군이 많이 없지만 미생물에 의해 발효된 된장이나 청국장에는 풍부해진다. 또 콩에는 비타민 C가 없지만 물을 주고 콩나물을 기르면 풍부해진다. 혈액 속에 호모시스테인이란 물질의 수치가 상승하면 관상동맥에 걸릴 가능성이 커지는데, 비타민 B_6과 B_{12}는 호모시스테인 수치를 낮추는 역할을 한다. 비타민 B군은 당류를 분해해 에너지를 만들 때 중요한 역할을 한다. 따라서 뇌를 활성화하려면 비타민 B군을 충분히 섭취해야 한다. 기억력과 관련된 중요한 영양소 중에서 중요한 것은 비타민 B_3(나이아신), 비타민 B_5(판토텐산), 비타민 B_6(피리독신), 비타민 B_{12}(코발라민)다. 특히 비타민 B_3은 기억력을 높이는데 큰 위력을 발휘한다. 비타민 B_{12}를 충분히 섭취하면 학습능력이 높아지고, 신경세포가 건강하게 성장하는데 도움이 된다. 비타민 B_6는 아미노산을 다른 아미노산으로 전환하는 역할을 한다. 이 작용으로 세로토닌이나 멜라토닌이 만들어진다. 비타민 B군은 서로 협력해서 뇌가 신경전달물질을 만드는 것을 돕는다. 따라서 B군은 같이 섭취하는 것이 바람직하다.

프로바이오틱스 효과

프로바이오틱스(probiotics)란 비피더스균, 유산균 등 우리 몸에 좋은 유익균에 대한 총칭이다. 대표적으로 좋은 균은 우리 몸속의

장내에 살고 있는 유산균인데, 장내의 나쁜 균과 싸우면서 항생물질을 분비하는 역할을 한다. 유산균은 처음 우유에서 분리해내서 생긴 이름이지만 된장과 청국장에도 많이 들어 있다. 된장의 유산균은 약 20%의 소금물에도 살 수 있는 내염성균이고, 섭씨 100도에서도 견뎌내는 내열성균이다. 최신양(한국식품개발연구원)에 따르면 된장의 추출물이 면역증강물질인 B 림프구(lymphocyte)의 증식을 증가시킨다는 것을 확인하였으며, 이러한 면역조절제가 삶은 콩에서는 발견되지 않고 된장에서 발견된다는 것은 자연계에 존재하는 다양한 미생물의 발효에 기인되는 것으로 알려져 있다.

갈색색소와 항산화작용

우리 몸 속에 들어온 활성산소는 세포막, DNA, 그 외의 모든 세포구조를 손상시키고 손상의 범위에 따라 세포가 기능을 잃거나 변질되기도 한다. 된장의 항산화효과는 된장의 갈색색소에 그 비밀이 숨겨져 있다. 된장이 발효하면서 생성되는 수용성 및 지용성 갈색색소가 항산화활성을 가져오기 때문이다. 이 색소가 우리 신체의 산화를 방지하여 노화를 효과적으로 예방해준다. 메주 및 된장의 발효숙성 중 갈변물질의 생성 등을 조사하고 이들의 갈변물질과 페놀화합물이 지질의 산화안정성에 미치는 영향을 검토한 결과, 메주 발효 중의 수용성 갈변물질은 함량은 낮았지만 원만한 증가를 보였고 된장 숙성 중에는 지속적으로 증가하여 높은 함량을 나타내었다. 메주 중의 페놀화합물 함량은 발효의 전반기에는 증가하였다가 후반기에는 다소 감소하였으나 수용성 갈변문질과 페놀화합물은 리놀레산에 대하여 강한 항산화효력을 나타내었다. 간장을 우려낸 된장의 추출물의 갈색도를 측정한 결과 숙성시간이 길어지면서

증가하였고, 지용성보다 수용성갈색물질이 더 큰 항산화효과가 있었다. 실험 결과 몸에 좋은 콩의 성분들을 대부분 그대로 가지고 있으면서 된장은 콩보다 기능성 물질의 함량이 높아졌다.

혈전용해효소

혈관에 쌓여 혈액의 흐름을 방해함으로써 심장병이나 뇌졸중의 원인이 되는 혈전을 녹여주는 혈전용해효소 등이 콩 발효식품에 함유된 대표적인 기능성 물질이다. 콩에는 혈전용해 효소가 전혀 없고 발효과정을 거친 된장, 청국장에서만 발견되고 있다. 이밖에 된장은 지질 산화과정 중 과산화물 생성을 억제하며, 혈소판 응집 저해 활성을 갖는 항혈전 펩타이드가 있는 것도 확인됐다.

면역증강물질

된장의 성분을 실험한 결과 된장의 추출물이 특이적으로 면연증강물질인 B림프구의 증식을 증가시킨다는 것이 확인되었다. 또한 대식세포와 B림프구의 싸이토카인 생성을 증가시켜 면역력 증강에 도움을 준다. 옛날엔 벌이나 뱀에 물리거나 하면 된장을 바르곤 했는데 된장이 염증과 관련된 유전자를 조절하는 기능이 있어 효험을 보기도 했을 것이다. 이러한 면역조절제가 삶은 콩에서는 발견되지 않고 된장에서만 발견되었다는 것에 주목할 필요가 있다. 즉, 삶은 콩은 가지고 있지 못한 성분이 된장에만 존재하는 것이다. 이 성분들은 된장의 발효과정 중에 관여한 다양한 미생물의 작용에 따른 콩의 기능물질이 면역기능을 높이는 물질로 전환하거나 발효산물에 의한 것으로 보인다.

콩과 궁합이 맞는 식품

모든 식품이 그렇듯이 한 가지만 먹어서는 인간이 하루에 필요로 하는 영양분이 충족되지 않는다. 콩이 아주 풍부한 영양분을 가진 것은 맞지만 완벽하지는 않다. 콩의 영양소를 보완하는, 소위 궁합이 맞는 좋은 식사법을 소개한다.

콩밥 : 콩 + 쌀

쌀은 전분이 많은 반면 단백질과 지질의 함량은 적다. 필수 아미노산 중에서는 리신 트립토판은 부족하고, 메티오닌은 많다. 이에 비해 콩은 전분이 적고, 단백질과 지방의 함량이 높은 편이다. 쌀은 콩에 부족한 메티오닌 함량이 많다. 쌀과 콩은 우리 밥상에서 오랫동안 서로의 단점을 보충해주는 궁합식품으로 짝을 이루어왔다. 떡에 콩을 넣어 먹거나 콩국수, 인절미를 해먹는 습관은 단백질 보충 효과에 도움이 된다.

콩국수 : 콩 + 밀

땀을 많이 흘리고 식욕이 떨어지는 여름철에 콩국은 최고의 건강식이다. 밀가루 중에는 단백질을 구성하는 필수 아미노산 중 라이신, 메치오닌, 트레오닌, 트립토판 등의 함량이 적은데 콩 단백질에는 이들 아미노산이 3~5배 정도 들어 있어 보완효과가 있으며 또한 콩에는 밀에 적게 함유된 비타민 B_1, B_2도 많이 들어 있다.

감두탕 : 감초 + 검은콩

감초는 비뇨기계와 폐기관을 보하고 기침을 멈추게 하는 효과가 있다. 또한 열을 내리고 독을 풀며 새 살이 잘 돋아나게 한다. 검은콩은 몸 안에 흡수된 독성물질을 쉽게 배설하도록 도와주거나 다른

물질로 분해시켜서 독성작용을 완화시켜준다. 감두탕은 중금속 해독에 효과가 큰 약용식으로 예로부터 이용되었다.

■ 된장찌개 : 된장 + 채소

된장 속에 들어 있는 나트륨성분을 보완해주는 것은 칼륨이 풍부한 식품이다. 된장을 끓일 때는 멸치와 다시마로 국물을 내고 부추, 감자, 호박, 표고 등의 채소를 넣어 먹는다. 특히 부추는 칼륨 함량이 높고 독특한 향내가 있으며 소화작용을 돕는 채소로 알려져 있다. 부추를 된장국에 넣고 끓여 먹으면 설사에 효력이 있다. 또한 부추는 장을 튼튼하게 하므로 몸이 찬 사람에게 매우 좋은 식품이다.

■ 미역국 : 해초 + 간장

우리는 생일날이나 산모가 해산하면 미역국을 먹는다. 아프리카의 어느 종족이 미역을 먹는다고는 하지만 출산 날, 또 해마다 돌아오는 생일날에 미역국을 먹는 나라는 지구상에 우리 밖에 없을 것 같다. 일설에는 바닷가에 살던 사람들이 고래가 산달이 되면 미역을 뜯어먹거나 상처에 미역을 감는 것을 보고 따라했다는 이야기가 있다. 콩에 함유된 사포닌은 항암 효과를 막아 주는 좋은 성분이지만, 체내에 들어오면 요오드를 몸 밖으로 배출하는 역할을 한다. 그러므로 항시 콩을 먹는 사람들에게는 요오드가 풍부한 식품 섭취를 하는 것이 좋다. 해조류에는 요오드가 풍부히 들어 있어 미역국에 간장으로 간을 해 먹는 방법은 매우 균형잡힌 식사법이라 할 수 있다. 미역국 외 미역초무침, 콩다시마조림, 두부해초무침, 다시마로 국물내는 된장찌개 등 다양하게 먹을 수 있다.

조한영의 『통속한의학원론』

1934년 조한영이 쓴 『통속한의학원론』에는 산후에 미역국 먹는 풍속에 대해 다음과 같이 설명했다. "조선에서는 산부에게 미역국(藿羹)을 될 수 있는 대로 다량으로 섭취시킨다. 본초학적 설명을 한다면 곽(藿)은 감미(甘味)와 담미(淡味)와 미약한 산미(酸味)를 가졌고 색(色)은 흑색과 청색과 황색을 대(帶)하여 미(味)로나 색(色)으로나 족삼음(足三陰) 즉 신경(腎經), 간경(肝經), 비경(脾經)을 배양(培養) 조정(調整)하는 약이다. 하여간 실제에 좋은 것만은 부인할 수 없는데 현대 의학상으로는 해조(海藻)는 대개 옥도(沃度, 요오드)를 함유했으며 특히 곽이 옥도의 함유량이 많다고 한다. 그런데 옥도는 다음과 같은 작용을 하여 산부에게는 가장 적당한 식물이 될 수 있다. (1) 신진대사 활발. (2) 병적 세포 내지 세균의 파괴 살멸. (3) 체내 독소의 중화 및 배출." 산후에 미역국을 먹는 처방은 기존의 의서와 본초서에 전혀 나오지 않는 우리 민족의 경험의학의 소산이다. 이에 대해서 이익(李瀷)의 『성호사설(星湖僿說)』에서는 다음과 같이 말한다. "월수(月水, 여자의 처음 나온 월경을 이름)가 열(熱)을 다스리고 미역(海藿)이 산부(産婦)의 선약이 된다는 것은 동방의 풍속에서 중요한 처방이라 하겠다."

콩제품을 이용한 생애주기별 건강밥상 만들기

우리는 현재 동물성 식품의 과잉섭취로 인해 증가되고 있는 서구형 질환을 해결하기 위해 국가적으로 많은 노력을 하고 있다. 그러나 그 해답은 아주 가까이 우리나라의 전통적인 식생활 소재에 있는 것이다. 콩은 우리나라에서 오랫동안 콩 자체, 장류로 발효시켜서, 또한 두부 등의 형태로 조리, 가공하여 섭취하여 왔다. 이러한 것의 섭취에 익숙하지 않은 서양인들과 달리 우리는 일상적인 우리의 콩식품의 섭취를 유지 또는 증가시키는 것만으로 해결 방안을 가지고 있는 것이다.[1)]

1) 승정자, 콩 음식의 영양가와 기능성, 한국콩박물관건립추진위원회편, 고려대학교출판부, 579-660(2005)

우리나라에서는 다양한 콩음식이 발달되어 왔다. 생애주기별로 비교해보면 총 열량은 비슷함에도 불구하고 지질 및 콜레스테롤의 섭취는 낮았고 식이섬유소, 칼슘, 이소플라본 함량은 높은 것으로 평가되었다. 국내 100세 장수인들을 조사한 보고서에 의하면, 장수식단은 우리 고유의 전통식단이며 건강한 장수인들의 공통된 특징은 콩 및 발효식품과 자연식품을 즐겨 먹는다는 점이다.

▪ 임신부 및 수유부 식단

콩제품으로 일부 식단을 대치하여도 수유부에게 증가되는 열량과 철분, 칼슘, 단백질공급이 가능하여 영양적으로 일반식단과 대등함을 알 수 있다.

- 콩밥, 미역국, 콩전, 멸치깻잎볶음, 콩나물부추나물, 나박김치

▪ 20~49세 및 50~64세 성인의 식단

동물성 단백질 대신 식물성 식품인 콩제품을 사용해도 열량 및 단백질 함량이 일반식단에 비해 감소되지 않으면서 성인병의 예방효과를 얻는다.

- 콩밥, 콩비지찌개, 삼색냉채, 도토리묵무침, 취나물된장무침, 열무김치

▪ 성장기 아동 및 청소년 식단

콩제품으로 대치한 식단으로 체조직 성장에 필요한 단백질을 충분히 공급해주며, 골 조직이 형성되는 중요한 시기이므로 필요한 칼슘이 일반식단에 비해 많이 함유되어 있다.

- 현미콩밥, 두부무국, 콩나물, 콩전, 오이양파무침

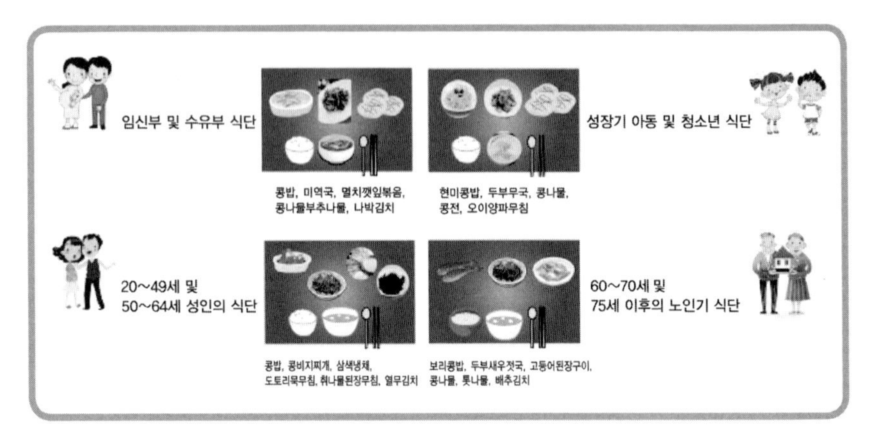

생애주기별 권장식단[1)]

■ 60~70세 및 75세 이후의 노인기 식단

단백질 및 비타민, 무기질 등의 영양소 섭취 부족으로 건강장애를 초래하는 경우가 많으므로 콩제품을 이용한 전통적인 식생활을 고려한다.

- 보리콩밥, 두부새우젓국, 고등어된장구이, 콩나물, 톳나물, 배추김치

4. 질환별 콩의 효능

콩에는 여러 기능성 물질들이 함유되어 있으므로 여러 질병의 예방과 치료에 사용되어 왔다. 옛날 사람들은 콩 및 콩발효제품을 민간의약품으로 사용하기도 하였다. 중국 명나라 때 나온 『본초강목』에는 "콩을 삶은 즙은 백약의 독을 푼다. 신장병을 다스리고

---

1) 승정자, 콩 음식의 영양가와 기능성, 콩, 한국콩박물관건립추진위원회편, 고려대학교출판부, 579-660(2005)

오줌을 이롭게 하며 기를 내린다. 모든 풍열을 억제하고 혈을 활발히 하며 모든 독을 푼다”라고 하였고 허준의 『동의보감』에서는 “벌레에 물렸을 때 검정콩 태운 가루를 물에 개어서 바르면 해독이 된다. 또 이 차는 오장과 위기의 결적을 치료하고 일체의 열독과 번갈 그리고 대·소변의 비삽을 치료한다”라고 하였다. 또 최근 집대성한 『향약대사전』에서도 “피의 흐름을 원활히 하고 이뇨작용을 하며 풍을 물리치고 해독의 기능이 있고, 황달부종, 수종창만 등을 치료하며 약독을 푼다”고 되어 있다. 검은콩식초가 통풍에 좋다는 말은 지금도 민간에 널리 퍼져있다.

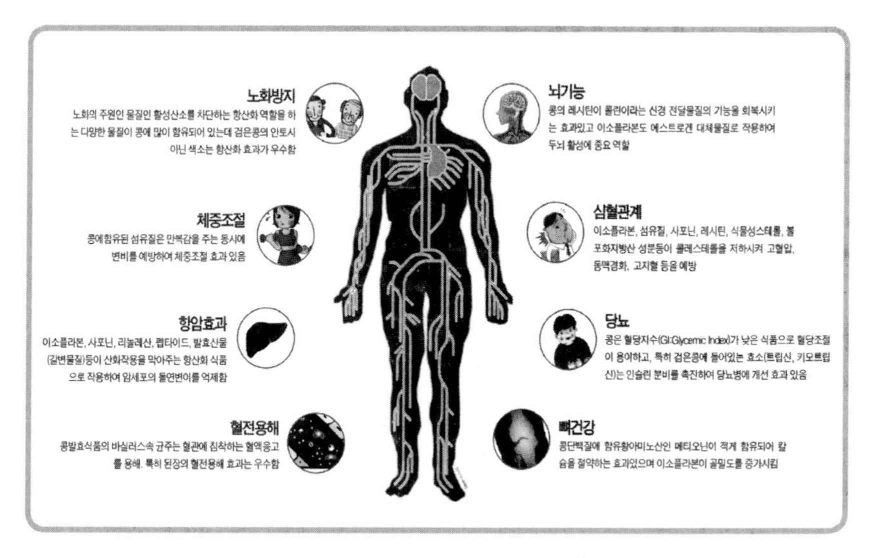

인체부위별 콩의 생리기능성[1)]

1) www.soyworld.org

질병의 예방과 치료에 유익한 콩

암에 좋은 콩

암세포는 정상세포가 여러 원인(세포의 산화, 운동부족, 스트레스 등)에 의해 돌연변이를 일으켜 비정상적인 세포로 된 상태로서 정상세포와는 다르게 세포분열을 계속하여 축적되고 기관의 구조를 변화시키며 다른 조직으로 전이하는 특징을 갖는다. 암의 전이는 신체 전체에 엄청난 손상을 가져오는 것이 보통이다. 예방법 중 하나는 산화작용을 막아주는 항산화식품을 섭취하는 것인데 콩이 대표적인 항산화식품이다. 콩 및 콩 발효식품의 성분 중 항암 효과가 있는 성분은 이소플라본을 비롯한 폴리페놀 화합물, 사포닌, 리놀레산, 토코페롤, 피트산, 펩타이드, 발효과정 중 형성된 갈변물질 등으로 특히 이소플라본의 항암성이 가장 주목받고 있다. 이소플라본은 여성호르몬인 에스트로겐과 화학적 구조가 비슷하여 에스트로겐 수용체와 결합할 수 있으므로 천연 에스트로겐이 수용체와 결합하여 유방세포 분화를 촉진시키는 것을 막아주는 원리로 작용한다. 또한 이소플라본의 하나인 제니스테인은 콩에만 함유된 물질로 전립선암의 진행을 저지하는 효과가 있다. 콩 발효식품 중에서 청국장의 항산화 효과가 특히 높은데 발효된 청국장에는 콩보다 8배 정도 많은 항산화 물질이 들어있다. 여러 실험을 통해 된장의 주 활성 물질인 제니스테인과 리놀레산은 암을 예방하는 작용이 매우 높은 것으로 보고되었다.

심혈관계질환과 콩

심장 순환계 질환은 심장 및 혈관에 이상이 생겨 혈액이 원활히 흐르지 못해 생기는 질병으로 고혈압, 동맥경화, 고지혈증, 심부전,

심근경색 등이 있다. 콩 이소플라본, 섬유질, 사포닌, 레시틴, 식물성 스테롤(phytosterol), 불포화 지방산, 펩타이드 등이 콜레스테롤을 저하시키고 심혈관 기능을 높이는 성분이다. 사포닌의 경우는 화학적 구조가 콜레스테롤과 유사하기 때문에 콜레스테롤 흡수를 저해하기도 하고 콜레스테롤 배출을 돕기도 한다. 생리활성 펩타이드[1]는 안지오텐신이라는 혈압을 높이는 화합물을 만들어내는 효소(안지오텐신 변환효소)의 기능을 방해하는 작용이 있다. 특히 청국장에는 단백질을 분해하는 강력한 프로테아제가 있어서 안지오텐신 변환효소 작용 저지효과가 우수한 식품이다. 대두단백질이 혈중지질 수준에 미치는 영향에 대한 연구결과를 종합 분석한 결과 하루 70~80mg의 이소플라본을 함유한 콩단백질(31~40g/day)섭취로 혈청콜레스테롤 수준이 유의적인 감소를 보였다.

혈전용해효과

혈전(fibrin clots)은 섬유의 일종인 피브린이 혈관 내에 침착한 것으로 일반적으로 인체의 혈관이 손상을 받아 출혈이 일어난 후 혈액이 혈관 또는 조직 내에서 응고됨으로써 생성된다. 혈전의 증상은 밤시간의 다리 경련, 다리가 무겁고 경직된 느낌, 장시간 서있거나 앉아있고 난 후 다리통증 등이다. 정상적인 경우는 혈전용해물질인 플라스민과 그 전구물질인 플라스미노겐에 의해 혈전의 형성과 파괴의 균형이 유지되지만, 비정상적인 경우는 혈액의 흐름이 제한되고 특정조직에 산소와 영양물질을 공급하지 못하는 혈전증(thrombosis)을 유발한다. 여러 문헌 조사 결과 청국장, 낫도, 된장 등의 콩 발효식품은 콩보다 혈전 용해 효과가 우수한 것으로 보고되

1) bioactive peptide; 특정 배열의 아미노산 사슬로서 2개-10개의 아미노산 사슬

었고 특히 된장에서 분리된 바실러스 속 균주는 낫도나 청국장에서 분리된 균주보다 혈전용해 효과가 3~4배 높은 것으로 조사되었으며 전통 된장의 혈전 용해능은 대량생산되는 된장보다 10배 높은 효과가 있는 것으로 보고되었다.

당뇨병과 콩

우리 몸의 위장 뒤에는 소화와 면역에 중요한 역할을 하는 췌장이라는 작은 장기가 있는데 이곳에서 혈당을 각 세포로 운반해주는 인슐린이 분비된다. 당뇨병이 무서운 이유는 당뇨로 인한 합병증이 나타나는 것으로 혈관에 문제가 생기기 때문에 신장, 심장, 눈 말초신경 등에 손상을 가져올 수 있다. 콩은 혈당지수(glycemic index)가 낮은 식품으로 혈당조절이 용이하고 가용성 섬유질을 함유하여 포도당이 혈액으로 들어가는 속도를 낮추고 세포들을 인슐린에 민감하게 만들어 주기 때문에 혈당 조절에 효과를 나타낸다. 콩에 함유된 피니톨은 인슐린 신호체계 작용을 도와 인슐린 유사작용을 하여 혈당 개선 효과를 나타낸다. 특히 검은콩에는 트립신과 키모트립신이라는 효소가 함유되어있어 이들이 췌장기능을 활성화하여 인슐린 분비를 촉진하기 때문에 당뇨병에 개선효과가 있다고 보고되었다. 문제는 콩에 함유된 효소는 열에 의해 활성을 잃게 되므로 (일반적으로 50℃ 전후가 효소활성의 한계온도로 보고됨) 검은콩에 함유된 효소의 혈당강하효과를 기대하려면 식초를 이용하여 절인다든가 하는 처리방법을 이용함이 바람직하다. 이외에 혈당을 저하시키는 콩의 성분으로 올리고당이 있다.

노화방지와 콩

생명체가 태어나 죽음을 맞기까지의 전 과정을 노화라 할 수 있는데, 노화의 주된 원인 중 하나는 활성산소(유해산소)이다. 우리 몸은 활성산소를 차단하는 방어시스템으로 항산화물질 SOD(Super Oxide Dismutase)를 만드는데 이들의 주요 역할은 세포내 대사과정에서 전자전달계에서 사용하고 넘치는 유리기(활성산소)가 다른 물질과 결합하여 산화되기 전에 이들과 결합하여 안정한 형태로 바꾸어 독성을 제거하는 것이다. 콩에는 항산화 역할을 하는 다양한 성분이 있는데 이소플라본, 안토시아닌등 페놀화합물, 트립신저해제, 아미노산, 펩타이드, 피트산, 비타민 A, C, E 등이 대표적인 물질이다. 안토시아닌은 플라보노이드계 색소로 과실이나 꽃의 홍색, 청색, 자색을 이루는 색소물질로 블랙푸드로 대표되는 식품 중에 많이 존재하며 검은콩의 항산화 효과는 메주콩의 4배로 조사되었다. 피트산은 금속성분 특히 철분과 쉽게 결합하는 성질이 있어 철분이 산소와 결합하여 유리기를 만드는 것을 방어함으로서 세포의 손상을 예방한다. 콩 발효식품은 이소플라본이 몸에 흡수되기 쉬운 형태로 변화되어있고, 발효 및 숙성되는 과정에서 수용성 갈변물질도 증가하여 콩과 비교할 때 8배 이상의 항산화 효과를 기대할 수 있다.

뇌기능(알츠하이머, 파킨슨)과 콩

인간의 두뇌 중 대뇌는 신체의 모든 활동을 지배하는 기관으로 무려 1,000억 개의 신경세포인 뉴런으로 이루어져있고, 사람은 태어나면서부터 생각하기 시작한다. 신경세포는 기억의 저장과 사용에 매우 중요한 역할을 하며 나이가 들면서 대뇌에 있는 신경세

포의 수는 점차 감소한다. 콩에는 레시틴이라 불리는 인지질이 함유되어 있는데 콜린이라는 신경전달물질의 기능을 회복시키는 효과가 있다. 알츠하이머형 치매의 뇌에서 감소하는 효소(콜린아세틸 트랜스퍼라제)의 활성은 에스트로겐 투여로 증가되는데 이소플라본이 에스트로겐과 비슷한 작용을 하여 두뇌 활성물질 중 기억력에 중요역할을 하는 에스트로겐의 대체 호르몬으로 작용할 수 있다.

뼈 건강과 콩

신체에서 새로운 뼈 조직이 형성되고 손상된 뼈 조직은 용해되는 복잡한 반응이 일어난다. 대표적인 뼈 질환으로 골다공증과 관절염이 있는데 보통 나이가 들어감에 따라 많이 발생한다. 동물성 단백질을 섭취하는 사람이 식물성 단백질을 섭취하는 사람들에 비해 골다공증에 걸리는 비율이 높은 것으로 보고되었는데 그 이유는 콩 단백질의 칼슘 절약효과로 설명된다. 콩 단백질은 함유황 아미노산인 메티오닌이 적게 들어있는 반면, 동물성 단백질에 많이 함유된 유황성분은 신장에서 칼슘이 재흡수 되는 것을 막아 많은 양의 칼슘이 소변으로 유실되도록 하는 작용이 있다. 함황아미노산은 체액을 산성으로 유도하여 소변으로 칼슘 배설량을 증가시키므로 칼슘의 손실을 막기 위해 콩 단백질을 식단에 첨가하는 것이 중요하다. 또한 콩에 함유된 이소플라본의 임상실험 결과 폐경기 이후 여성의 뼈 밀도를 증가시킬 뿐 아니라 관절염의 원인이 되는 바람직하지 않은 혈관신생작용을 억제하는 효과도 있다는 것이 보고되었다.

체중조절과 콩

비만은 여러 퇴행성 질환과 상관관계가 큰 질병으로 지방섭취량의 증가 및 생활양식 등의 변화로 해마다 증가하는 추세로 건강을 위해서는 정상체중을 유지하려는 노력이 중요하다. 섭취한 칼로리보다 소모한 칼로리가 적은 경우 또는 칼로리 섭취량과 소모량은 균형을 이루는데 칼로리 대사에 이상이 생기는 경우 비만이 된다. 비만의 치료는 상당히 오랜 시간이 걸리므로 인내심을 가지고 식이요법과 운동요법, 행동수정요법 등을 병행해야한다. 섬유질은 만복감을 줌과 동시에 변비를 예방하여 체중조절 효과를 기대할 수 있으며, 불포화지방산은 중성지질을 감소시켜 체지방 조절에 효과가 있으므로 콩이야말로 저칼로리, 저지방, 최적의 다이어트 식품인 셈이다.

Chapter 06

콩의 미래

1. 콩과식물과 질소고정
2. 환경을 살리는 콩제품
3. 아프가니스탄에서 희망을 본다
4. 야생콩 vs 유전자변형(GM)콩
5. 콩의 생산과 교역 현황
6. 한국콩의 미래 전망

1. 콩과식물과 질소고정

근류균의 발견

일찍이 동양에서는 콩을 심으면 땅이 기름지게 된다는 것을 알고 있었다. 하지만 콩을 뒤늦게 접한 서양에서는 근류균이 발견될 때까지, 농업에서 콩이 얼마나 중요한 역할을 하는지 모르고 있었다. 1885년 토양 미생물학자인 네덜란드의 M.W. 바이에링크는 질소고정을 담당하는 뿌리혹박테리아를 분리해내었다. 뿌리혹박테리아는 뿌리의 조직을 군데군데 크고 뚱뚱하게 혹처럼 만드는 세균으로 '공생유리질소고정균' 혹은 '근류균'이라고도 한다. 근류균은 고등식물과 공생하여 유리질소를 고정하는데 식물 종류에 따라 균의 계통이 달랐다. 많은

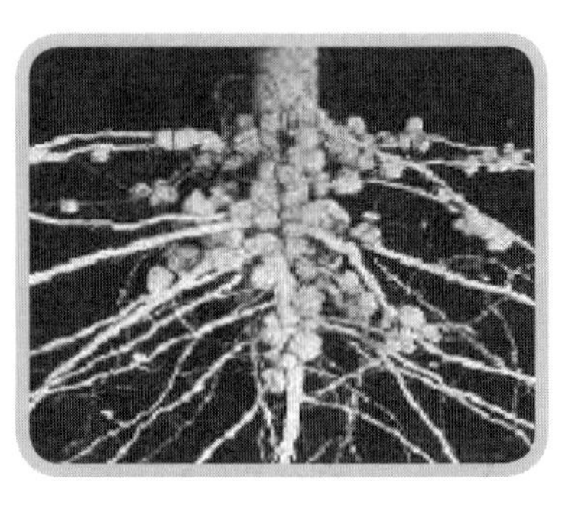

식물종에서는 근류균이 기생하지 못한다. 공기 중에는 78% 정도의 질소가 있긴 하지만, 보통의 식물은 공기 중의 질소를 거의 이용하지 못한다. 하지만 콩의 뿌리에 있는 근류균은 공기 중의 질소를 암모니아로 바꾸는 능력이 있다. 암모니아는 효소에 의해 아미노산이 되고 이어 단백질로 합성이 된다. 바로 이 근류균 덕분에 콩은 40%의 단백질을 함유하게 되는 군계일학의 식물이 되는 것이다.

콩과식물의 구분

콩과(Leguminosae)는 전 세계, 특히 열대지방에 약 550속, 13,000종이 있으며 우리나라에는 36속 92종이 분포되어 있다. 콩(大豆)은 콩과에 속하는 한해살이 초본식물로서 학명은 *Glycine max (L.) Merill*이다. 콩은 발아할 때 한 개의 어린 뿌리가 자라서 곧은 뿌리의 원뿌리가 되며 여기에서 많은 곁뿌리가 발생하고 뿌리에는 많은 뿌리혹이 착생한다. 뿌리혹 속의 뿌리혹박테리아는 공중질소를 고정하여, 콩 식물체가 바로 이용할 수 있도록 질소화합물을 만든다. 즉 콩과식물은 천연질소공장인 셈이다. 콩과(科) 식물은 씨의 모양과 싹이 난 묘종의 형태로 속(屬)이나 종(種)을 구분할 수 있다. 씨는 그 모양과 배아의 길이, 모양, 높이 및 위치를 보고 분류가 가능하다. 싹이 난 후에 떡잎의 위치가 지하에 있는가 지상에 있는가, 제1엽에 잎자루가 있는가 없는가, 잎 밑동의 형태 및 떡잎의 모양과 수를 보고 구별한다. 지하성 떡잎을 가진 종은 완두, 팥, 잠두, 비둘기콩, 날개콩 등에서 볼 수 있다. 이것들은 밑에 하배축이 자라지 않기 때문에 지상으로 올라오지 못한다. 콩과의 잎은 대부분이 겹잎이지만 제1엽은 홑잎이고, 대개는 제3엽에서 겹잎으로 바뀐다.

질소의 순환

질소는 모든 동물과 식물에게 필수적인 물질로 사람의 경우 하루에 16g 이상의 질소를 필요로 한다(공기의 성분: 질소 78% > 산소 21% > 아르곤 0.93% > 이산화탄소 0.04%). 그러나 공기 중에 있는 질소기체는 우리가 호흡하는 동안 우리 몸을 들락거릴 뿐 일체 인체 내 생화학적 반응에 참여하지 않는다. 왜냐하면 질소분자(N_2)는 섭씨 1,000도 이상으로 가열해야 끊어질 정도로 단단하게 결합되어 있기 때문이다. 고정화된 질소는 식물이 흡수해서 단백질이 된다. 동물이 식물을 먹으면서 이 단백질도 함께 먹게 된다. 그후 동물 몸속의 배설물을 통해 암모니아나 요소의 형태로 다시 땅으로 돌아가거나 질소 분해로 대기로 돌아가기도 한다. 콩과식물은 질소 순환에 큰 역할을 한다. 콩과식물에 기생하는 뿌리혹박테리아는 공기 중의 질소를 고정시켜 콩과식물들이 이용할 수 있도록 할뿐만 아니라 토양 중에 남아있는 질소는 다음 작물이 이용하게 한다.

질소고정(Nitrogen fixation)

질소고정이란 대기 중의 질소(N_2)가 암모니아(NH_3)로 전환되는 것을 말한다. 대기 중의 질소는 질소-질소간 강한 3중 결합을 가지고 있어 화학적으로 불활성에 가깝다. 다른 화학물질과 반응을 하거나 관계를 주고 받기 어렵다는 의미이다. 그러므로 질소고정이란 화학적으로 불활성인 질소기체를 활성인 질산염의 형태(암모니아)로 바꾸는 과정이라고도 할 수 있다. 질소고정은 자연적으로 일어나지만 산업적으로 합성되어 이용된다. 번개에 의해 물리적으로도 질소고정이 일어나기도 하지만, 우리가 질소고정을 이야기할 때 대부분은 미생물에 의해 일어나는 생물학적 질소고정을 일컫는다. 콩 뿌리

에 공생하는 리조비움 박테리아는 연간 헥타르(ha)당 100~400kg의 질소를 고정하여 콩에 공급하여 주는데 이는 시판되는 요소비료 43포대와 같은 양이 된다.

콩과식물과 근류균의 공생

대부분의 식물들은 토양 내에 포함된 질소성분의 양에 따라 성장이 조절된다. 하지만 예외적으로 콩과식물들은 리조비움(Rhizobium)이라 불리는 토양 박테리아와 공생관계를 유지함으로써 이런 제약조건으로 부터 자유로울 수 있다. 리조비움은 공기 중의 질소성분을 고정시킬 수 있는 박테리아로 숙주 식물에 유입되면 식물의 뿌리에 작은 혹(nodule)을 만들어 그 안에서 살아간다. 이와 같이 서로 다른 두 종이 밀접한 관계를 유지하면서 서로 이익을 얻는 것이 공생(symbiosis)이다. 식물은 박테리아의 생존에 필요한 여러 가지 영양분들을 공급하고, 박테리아는 고정한 질소성분을 식물에게 제공한다. 콩은 뿌리에 기생하는 근류균과의 공생관계를 통해 공기 중의 질소를 고정하여 이용하는 대표적인 콩과식물이다.

2. 환경을 살리는 콩 제품

콩은 한마디로 '지구의 어머니'같은 식물이라고 할 수 있다. 토양 내의 질소를 고정해 단백질이 많은 식물, 즉 콩으로 자라나 토양을 기름지게 하고 인간에게는 고분자 단백질을 제공해준다. 자연적으로 질소가 순환되는 방법은 크게 3가지인데 첫째는 식물이나 동물이 썩거나 죽어 질소원이 되는 방법, 둘째는 번개에 의한 질소의 고정, 셋째는 콩 및 콩과식물의 뿌리혹에서 이루어지는 질소고정이다. 이것이 인공질소가 발명되기 전의 모습이다. 20세기 초 화학적

인 방법으로 생산된 질소를 사용하여 기하급수적으로 늘어난 인구를 부양해온지, 채 1세기가 지나지 않아 문제점이 드러났다. 화학비료(인공질소)의 남용으로 수질악화, 즉 수질의 부영양화(富營養化)가 초래되어, 오히려 생산성이 떨어지고, 산소고갈과 대량폐사의 위험도 떠안게 되었다. 더구나 화학비료의 생산은 석유 에너지를 기반으로 하고 있어 석유의 고갈이 예견되고 있는 최근에는 콩과작물의 친환경적 가치가 커지고 있는 실정이다. 홍은희 박사에 의하면, 콩과작물은 연간 ha당 수백kg의 질소를 고정하여 지력유지 및 환경보전 효과가 크고, ha당 연간 3톤의 CO_2를 고정하고 산소는 2.18톤을 방출하여 대기 정화능력이 크다. 또한 콩을 많이 재배하게 되면 토양유실 경감효과도 있어 경사지 7% 이상의 밭 토양 유실량을 50% 이하로 방지한다.[1)]

친환경제품의 증가

최근에 친환경적인 제품에 관한 인식이 높아지면서 콩 제품의 사용 범위와 사용량이 계속해서 증가하고 있다. 콩은 21세기 환경문제를 해결하고 석유자원을 대체할 친환경 대체산업재로서의 가능성도 무궁무진하다. 콩은 다양한 분야들에 응용 적용되고 있으며 그 적용분야는 더욱 커질 것으로 예상되고 있다. 콩을 이용한 친환경 제품은 1970년대 초 원유가 가 급등하면서 미국에서 개발되기 시작하였다.

콩기름 잉크

원유와 비교해 원재료 비용도 비싸지 않고 공해가 없는 천연제품

---

1) 홍은희, 콩 재배 역사, 콩, 한국콩박물관건립추진위원회편, 고려대학교출판부, 103-136(2005)

으로 환경적인 차원에서도 권장될 만하다. 콩 인쇄잉크의 장점은 재생산이 가능한 원료를 사용하고, 생생한 색상을 가지고 있으며, 휘발성 유기물질이 낮아 폐신문지 재활용시 시간이 단축되며, 특히 신문을 볼 때 손에 묻지 않는다. 또 사용 후 잉크 분리가 쉬워 종이의 재활용을 쉽게 한다.

이소플라본 화장품

콩의 기능성물질인 이소플라본이 화장품에도 쓰인다. 항암, 항산화작용, DNA노화억제, 활성산소 억제기능이 화장품에 고스란히 담겨, 피부 재생과 주름개선에 효과적이다.

콩으로 만든 식물성 세제

생분해도가 99.9%에 달한다. 세정력도 뛰어나서 때가 씻겨나가는 모습이 눈으로 확인된다. 콩에서 단백질을 추출하여 만든다.

콩섬유

콩단백질을 반죽하고, 이를 소금과 산이 담겨있는 응고조에서 응고시켜 실을 만든다. 이렇게 만들어진 실로 옷감을 만들고 염색, 가공하여 옷을 만든다. 콩섬유는 가볍고 부드러워 식물성 캐시미어라고 불리며 통풍성이 높고, 생산공정이 환경에 해를 주지 않아서 친환경 건강섬유라고도 불린다. 콩의 주성분인 콩단백질로 만드는 것으로 자연생태계 속에서 저절로 분해된다. 콩이 갖는 안정성 때문에 식품 포장용으로 아주 적합하고 사용한 뒤에는 흙 속에서 분해되어 땅을 비옥하게 만들고 분해 속도가 다른 석유화학 제품과는 비교가 안 될 만큼 빠르며, 한편으로 분말로 만들어 가축의 단백질 사료로도 재사용이 가능하다.

▪ 콩으로 만든 건축자재

콩가루 40%, 재활용된 신문지 40%, 다른 재료 20%로 이루어졌다. 이 자재는 견고한 목재 같으면서 대리석 같은 느낌을 준다. 고품격의 마루, 장롱, 가구, 장식벽 등의 재료로 좋다고 한다.

▪ 바이오 솔벤트

기름, 접착제, 페인트 등을 제거할 때 쓴다. 무독성으로 석유물질로 인한 오염제거 작업, 토양오염제거 작업 등에 사용한다. 또한 소이왁스로 만든 소이캔들(soy candle)은 그을음이 적고 깨끗하게 연소되어 유해물질을 남기지 않는 장점이 있다.

▪ 바이오디젤

바이오디젤(biodiesel, biofuel)이란 폐식용유, 콩기름 등의 식물성 기름을 알코올에 반응시켜 만드는 것으로 산소를 포함하고 있는 친환경 제품이다. 식용으로 사용되는 식물성오일이기 때문에 인체에 무해하며, 자연계에서 28일 경과 시 77% 이상 분해되는 특징을 가지고 있다. 특히 기후 변화 협약과 관련하여 바이오디젤의 1톤 사용시에 온실가스인 CO_2를 2.2톤 감소하여 주기 때문에 향후 전세계적으로 보급이 확대되리라 전망된다. 콩은 이러한 바이오디젤 생산을 위해 떠오르는 원료물질이다. 우리나라에서도 2002년 5월부터 자동차 경유의 대체품인 바이오디젤 혼합유의 시범보급이 이루어지고 있다.

3. 아프가니스탄에서 희망을 본다

화석연료를 사용하여 소고기에서 1칼로리의 단백질을 생산하기 위해서는 화석연료 54칼로리를 태워야 하는 반면, 콩에서 1칼로리

의 단백질을 생산하기 위해서는 2칼로리만 태우면 된다. 다시 말하면 콩에서 단백질을 섭취하는 사람의 에너지 소비는 고기에서 섭취하는 사람의 4%에 불과하고, 콩에서 단백질을 섭취하는 사람의 이산화탄소 발생량은 고기에서 섭취하는 사람의 4%에 불과하다는 의미이다. 이산화탄소 다음으로 지구 기후에 불안요소로 작용하는 가스는 메탄이다. 메탄은 사실 이산화탄소의 24배에 달하는 온실효과를 발생시키면서, 대기 중 농도도 훨씬 빨리 상승시키는 성질이 있다. 미국 환경보호청에 따르면 전 세계 대기 중에 떠 있는 인간에 의해 생성된 메탄가스 중 25%는 목축업 때문에 생성된 것이라고 한다. 1파운드의 소고기를 생산할 때 환경에 미치는 피해가 1파운드의 파스타를 생산할 때의 20배에 달한다고 결론지었다. 인간과 식품, 지구의 관계를 새로 정립하는 것을 하나의 역사적 혁명이라고 하였다. 환경 생태계와 식량의 위기는 인류가 지속적으로 생존하는 데에 가장 절박한 문제임이 더욱 명확해졌다. 전 지구적으로 갈수록 더욱 심화되어가고 있는 생태계의 위기는 이제 인류의 생존자체를 직접적으로 위협하고 있기 때문이다.

브라질의 경우

"브라질에서 열대우림 파괴의 주범은 가축 사육이다. 우리는 열대우림이 햄버거로 바뀌는 장면을 보고 있는 것이다." (노먼 마이어스, 가장 중요한 자원 : 열대우림과 인간의 미래) 브라질은 20세기 후반 들어 세계 제 1위의 소고기 수출국이면서 세계 제 2의 콩 생산 대국으로 성장했다. 아마존 밀림 파괴의 주범은 대규모 목장과 대규모 콩밭 조성이다. 브라질의 콩은 마토 그라소주를 위시한 중서부지역의 대규모 개간농장에서 생산되는데 동부 해안 산토스

항에만 매일 4톤 트럭 200대 분량의 콩이 생산되고 있다. 이렇게 대량 생산되는 브라질의 콩은 그러나 25%의 콩기름을 제외하고는 대두박의 형태로 수출되어 결국은 소와 같은 가축의 사료로 사용되고 있다. 브라질은 미국에 이어 콩의 생산량이 급격하게 늘어나고 있는 나라다. 지구의 허파라고 불리는 아마존의 밀림지역이 콩을 심기 위해 파괴되고 있다. 일견 보면 식량을 심으니 좋은 일이라 하겠지만, 결국은 브라질의 콩이 미국의 사료공장으로 직행한다는 점에서 시사하는 바가 크다.

아프가니스탄의 경우

아프가니스탄에서는 2006~2008년까지는 15개 주, 2009년에는 11개의 주, 2010년에는 34개 주 모두에서 콩이 생산되는 기적이 일어나고 있다. 네슬레의 영양담당 임원으로 근무하던 권순영 박사는 어느 날 아프가니스탄에 갔다가 많은 산모들이 아이를 낳다가 영양실조로 죽어간다는 사실을 알게 되었다. 물론 영유아의 사망률도 세계 최고 수준이었다. 권박사에게 떠오른 해결책이 있었으니 바로 콩이었다. 콩은 가격도 저렴한데다가 콩의 단백질과 칼슘을 공급하면 산모나 영유아의 건강이 좋아지리라고 확신했다. 그는 재원이 가능한대로 두유를 만들어 영유아들에게 먹이기 시작했다. 아이들은 놀라우리만치 빠르게 건강이 회복되었다. 그는 점점 두유의 양을 늘리고 한편으로는 그들의 주식인 난(Nan)을 만들 때 10%의 콩가루를 섞어 만들었다. 마침내 그는 NEI(영양과 교육인터내셔널)을 만들어 아프가니스탄에 콩을 직접 재배하기 시작했다. 양귀비를 키우던 나라에서 7년 만에 아프가니스탄의 34개주 전 지역으로 콩 재배지역이 확대되었다.

4. 야생콩과 유전자변형(GM)콩

야생콩의 이용

우리나라의 산야에 지천으로 자라고 있는 야생콩은 재배콩의 조상식물이다. 재배콩은 우리 땅에서 오랜 기간 동안 야생콩으로부터 진화해 왔다. 야생콩은 기능성이나 특정 병해충에 대한 저항성, 재배적 특성 등 유용한 형질을 다수 보유하고 있다. 콩은 우리 민족이 오랫동안 식용해온 된장, 간장, 콩나물 등 전통식품의 주원료인 만큼 친환경농업과 식량자급, 식량안보 차원에서도 콩의 생산자급기반 확충은 필요하다. 최근엔 수입 GM콩에 대한 소비자 불신으로 국산콩의 안정성과 우수성(장류적성 등)이 강조되어 국산콩 수요가 늘고 있으며 이소플라본, 안토시아닌 등 고기능성 콩에 대한 수요도 꾸준히 증가되고 있다. 한편 우리나라는 콩의 원산지국가로서 또 종주국으로서 다양한 야생콩의 유전자원을 활용하여 부가가치가 높은 콩을 개발해야 할 의무가 있겠다.

기능성 콩 품종의 육성 사례

주요 식량작물 중 우리나라가 원산지인 작물은 콩밖에 없다. 콩 종자는 40%가 단백질이고 지방이 20%나 되어 다른 어떤 작물 종자보다도 높은 영양가를 가지고 있다. 야생콩과 재배콩은 염색체의 수도 같고 서로 간에 교잡이 가능하다. 야생콩의 유용형질과 기능성을 가진 새로운 콩 품종을 만들기 위해서는 오랜 기간 동안의 연구와 실험이 지속적으로 이루어져야 한다.

이소플라본 함량이 높은 품종의 육성

재배콩과 야생콩 간의 종간교잡을 통해 이소플라본의 함량이

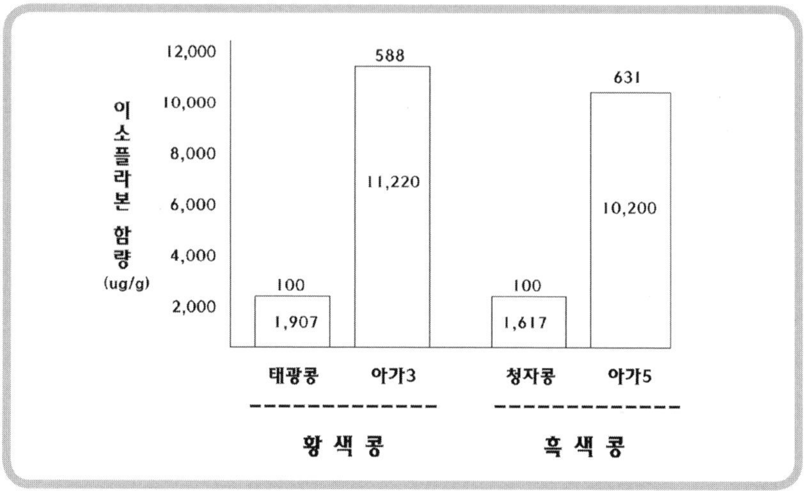

이소플라본 함량 (ug/g)
12,000
10,000
8,000
6,000
4,000
2,000
태광콩 1,907 (100)
아가3 11,220 (588)
청자콩 1,617 (100)
아가5 10,200 (631)
황색콩
흑색콩

아가콩의 이소플라본 함량 비교

지금까지 알려진 세계 최고인 기능성 콩 품종(아가콩)이 개발되었다. 고이소플라본 콩 첨가는 이소플라본 함량이 10,000ug/g 내외로 우리나라 황색콩과 흑색콩의 대표적 품종인 '태광콩'과 '청자콩' 보다 이소플라본 함량이 약 6배나 높다.

사료용 콩 품종의 육성

가축의 사료로 가장 많이 이용되는 작물은 옥수수이다. 옥수수는 탄수화물의 함량이 높은 작물로서 가축에 필요한 단백질과 지방의 함량은 낮다. 야생콩과 재배콩 간에 인공 교잡된 덩굴성의 새로운 콩 품종을 육성하고 옥수수와 혼작 후 함께 수확하면 전체사료의 단백질이 옥수수 단작보다 40%, 지방이 30% 이상 증가되어 양질의 사료가 된다.

고 올레산 콩 품종의 육성

양질의 식용유로 알려진 올리브유의 혈압강하 기능성은 약 75%를 차지하는 올레산에 의한 것이다. 2014년 우리나라에서는 전통적인 육종방법으로 일반콩에 비해 올레산의 함량이 4배나 높은 기능성 콩을 육성하였다. 새로 육성된 고 올레산 콩 품종의 오메가3(리놀렌산)와 오메가6(리놀레산)의 비율은 1:1이다. 하지만 야생콩을 이용하면 오메가3과 오메가6의 함량비가 2~3:1로 높아진다.

고 올레산콩 품종의 지방산 함량과 비율

종 류	올레산(%)	오메가3 : 오메가6
올리브	75	-
일반 콩 품종	23	6~7:1
고 올레산 콩 품종 (재배콩 × 재배콩)	80	1:1
고 올레산 콩 품종 (재배콩 × 야생콩)	70	2~3:1

유전자 변형(GM) 콩의 개발과 이용

유전자 변형에 의한 생명공학기술은 지속적인 인구증가와 경지면적의 감소에 따른 식량문제 해결, 식품의 기능성 강화, 에너지원 개발, 환경문제, 난치병 치료 등의 대안으로 개발되었다. GM(Genetically Modified)작물은 해충관리(미생물농약), 잡초관리(제초제내성), 질병관리(바이러스, 박테리아 저항성), 기능성 강화(비타민 등) 식량증산의 목적으로 분자육종기술을 이용해 만든 신품종들이다. 제초제내성GM콩은 박테리아에서 분리한 제초제 내성유전자가 삽입된 유전자변형콩으로 1996년 상업화에 성공하였는데, 이 GM콩은 제초제에 견디는 힘이 있어 제초제를 치면 잡초는 죽지만 GM콩

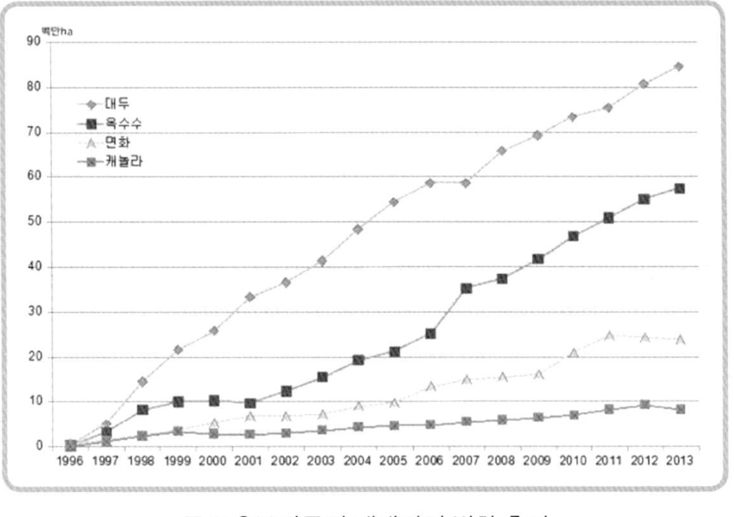

주요 GM작물의 재배면적 변화 추이

은 견디기 때문에 콩밭의 풀을 쉽게 없앨 수 있는 장점이 있다. GM작물은 철저한 안전성 시험과 평가를 거쳐 안전성이 확인된 것만 재배가 허용되며, 수입국에서도 다시 안전성 평가를 하여 허가하고 유통하도록 하고 있다. GM작물 중에서 콩의 재배면적이 가장 가파르게 증가하고 있다. 이것은 콩을 주로 수출용으로 재배하는 미국, 브라질, 아르헨티나 등 대규모 영농에서 제초제 사용이 불가피하기 때문이다.2)

GMO의 안정성 논란과 표시제도

유전자변형생명체(GMO)의 안전성에 대한 논란은 제초제내성 GM콩과 해충저항성(Bt) GM옥수수가 상업적으로 재배되면서 본격화 되었다. 그 동안 일부환경 단체나 소비자 단체들의 집요한

2) 박수철, 김해영, 이철호, GMO 바로알기, 도서출한 식안연(2015)

안전성 논란에도 불구하고 대규모 영농을 하는 미국, 브라질, 아르헨티나, 호주 등 주요 곡물 수출국에서 GM작물 재배가 급격히 증가하고 있다. 2014년 통계에 의하면 전 세계 28개국, 1억 8,150만 헥타르에서 GM작물이 재배되고 있다. 전 세계 콩 재배 면적이 79%에서 GM콩이 재배되고 있으며, 미국에서 생산되는 콩의 90% 이상이 GM콩이다. 따라서 생명공학기술에 의한 GM작물의 생산은 세계적인 대세이며 앞으로 인류의 식량문제를 해결할 중요한 기술로 인식되고 있다.

GMO표시제도는 나라마다 자국의 형편에 따라 다르게 운영되고 있다. GM작물개발과 이용에 선도적인 역할을 하는 미국은 실질적 동등성(substantial equivalence) 원리에 근거하여 일반작물과 GM작물이 성분 조성에서 차이가 없으므로 아무런 표시를 하지 않고 사용하고 있다. 반면, 외국의 값싼 곡물 수입을 막아 자국의 농업을 보호해야 하는 유럽 국가들은 GMO표시를 까다롭게 하여 모든 GM식품의 표시를 의무화 하고 있다. 식량의 대부분을 수입에 의존하고 있는 한국, 일본, 대만 등지에서는 GM식품을 표시하되, 가공과정에서 외래 유전자나 그로 인한 단백질이 잔류하지 않아 식별이 불가능한 식용유, 간장, 물엿 등은 표시하지 않아도 된다.

5. 콩의 생산과 교역 현황

2013/2014년 콩의 세계 총 생산량은 2억 8,530만 톤이었으며 이중 39.5%에 해당하는 1억 1,283만 톤이 다른 나라로 수출되었다. 콩의 5대 수출국, 브라질, 미국, 아르헨티나, 파라과이, 캐나다는 모두 GM콩 생산국이며 각국의 2014년도 GM콩 생산비율(채택율)은 93%, 94%, 100%, 97%, 95%이었다. 이들 국가에서 수출된 콩의 양은

콩의 세계 생산량, 주요 생산국, 수출국 및 수입국(2013/2014)

단위: 1,000톤

순위	총생산량 : 285,302 / 교역량(39.5%) : 112,829(수출량 기준)					
	7대 생산국		5대 수출국		10대 수입국	
	국가	생산량	국가	수출량	국가	수입량
1	미국	91,389	브라질*	46,829	중국	70,364
2	브라질	86,700	미국*	44,815	EU	12,950
3	아르헨티나	54,000	아르헨티나*	7,841	멕시코	3,700
4	중국	12,200	파라과이*	4,400	일본	2,894
5	인도	11,000	캐나다*	3,471	대만	2,350
6	파라과이	8,200			인도네시아	2,200
7	캐나다	5,359			태국	1,798
8					이집트	1,674
9					터키	1,608
10					베트남	1,350

*주요 GM콩 생산국의 수출량 합계 107,356, 자료: USDA(2013/14)

1억 753만 톤으로 전체 교역량의 95.1%에 달한다. 각국의 GM콩 채택률에 비례하여 GM콩이 수출되었다고 가정하면 총 GM콩 교역량은 1억 108만 톤이며 전체의 90%에 해당한다.[1)]

우리나라의 콩 수급 현황

한국은 1970년도까지만 해도 콩의 대부분을 자급하였다. 전체 콩 자급률은 86.1%, 식용 콩 자급률은 92.3%에 달했다. 물론 이시기는 절대빈곤의 시대였으므로 일인당 콩 소비량은 연간 5.3kg 수준으로 충분한 양은 아니었다. 경제성장과 더불어 동물성 식품의 소비가 늘고 이에 따라 축산업이 장려되면서 사료용 콩 수요가 급속히 늘어 1980년도 콩 자급률은 35%, 1990년에는 20%로 급감하였다. 우루

1) 박수철, 김해영, 이철호, GMO 바로알기, 도서출판 식안연(2015)

우리나라 콩 수급 및 자급률

연도	1970	1980	1990	2000	2013
생산량(천 톤)	232	216	233	113	123
수입량(천 톤)	36	417	1,092	1,567	1,153
식용 콩 자급률(%)	92.3	64.3	64.9	28.2	29.1
콩 자급률(사료 포함, %)	86.1	35.1	20.1	6.4	9.7
1인당 연간 소비량(kg)	5.3	8.0	8.3	8.5	8.0

자료: 선진국의 조건 식량자급, 2014, 농림축산식품 주요통계, 2014

과이라운드(UR)협상이 끝나 WTO 무역자유화가 되면서 콩의 자급률은 급격히 떨어져 10%를 밑돌게 되었다.[1)] 수입콩의 대부분은 사료용으로 사용되고 있다. 2010년도 우리나라 전체 콩 수요량 159만 톤 중 사료용으로 사용된 콩이 117만 톤(74%)이었으며 식용으로 사용된 양은 전체 콩 수요의 26%에 불과하다. 국산 콩은 거의 전량 식용으로 사용되고 있으며 부족되는 부분을 수입콩으로 충당하고 있다. 2008년부터 3년간 국내 평균 식용 콩 소비량은 연간 40만 6,000톤인데 반해 생산량은 12만 6,000톤으로 평균 식용 콩 자급률은 31%를 기록하고 있다.[1)] 우리 식단에서 콩의 중요도를 생각하면 식량안보적 차원에서 식용 콩만이라도 자급하려는 노력을 해야 한다.

콩의 용도별 수입현황

대두유와 경화유 제조에 사용되는 착유용 대두와 사료용 콩은 모두 수입 GM콩이다. 콩 가공제품으로 수입되는 사료용 대두박, 제과 제빵에 사용되는 경화유와 2차 유지가공 제품들은 기타로 분류되며 모두 GM콩으로 만들어 진다. 기타로 분류된 수입 대두박

1) 이철호, 문헌팔, 김세권, 이숙종, 이꽃임, 선진국의 조건 식량자급, 도서출판 식안연 (2014)

과 경화유 등 2차 유지가공제품의 수입량은 전체 두류 수입량의 1/2 수준에 달하며 2012년도 수입량은 153.9만 톤이다. 식용으로 사용되는 콩나물콩과 두부와 장류 제조에 사용되는 콩은 non-GM 콩이다. 2012년 착유용과 사료용으로 사용되는 GM콩 수입량은 115.2만 톤이며, 식용으로 사용되는 non-GM콩의 수입량은 31.9만 톤으로 전체의 28%이었다. 2012년도 국산콩 생산량은 12.9만 톤으로 국내 non-GM콩 공급량은 44.8만 톤으로 추산된다.[1] 2012년 GM콩 수입액은 7.3억 달러이며 non-GM콩 수입액은 2,5억 달러이다. Non-GM콩의 수입액 비중은 전체의 22%이다. 수입물량 비중보다 수입액 비중이 낮은 것은 식용으로 들어오는 수입 콩 대부분이 저율관세할당물량(TRQ)으로 들어오기 때문이라고 생각된다.

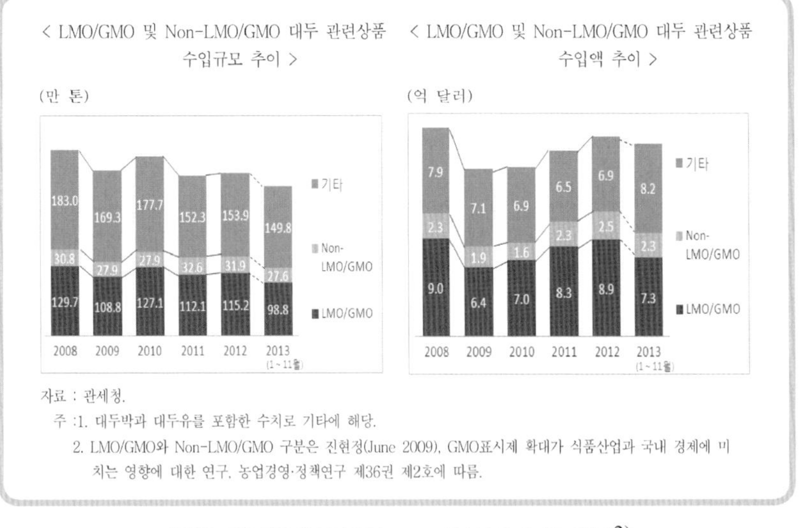

자료 : 관세청.
주 :1. 대두박과 대두유를 포함한 수치로 기타에 해당.
2. LMO/GMO와 Non-LMO/GMO 구분은 진현정(June 2009), GMO표시제 확대가 식품산업과 국내 경제에 미치는 영향에 대한 연구. 농업경영·정책연구 제36권 제2호에 따름.

콩의 용도별 GM콩과 non-GM콩 수입 현황[2]

1) 박수철, 김해영, 이철호, GMO 바로알기, 도서출판 식안연(2015)
2) 이부형 외, 국내외 주요 곡물 수급 환경변화 전망에 관한 연구, 현대경제연구원(2014)

6. 한국 콩의 미래 전망

문자가 사용되기 이전 한반도와 남만주에 웅거했던 동이족에 의해 처음 식용으로 사용되기 시작한 콩이 이제 전 세계인의 주요 식량이 되었고 앞으로 그 영양 기능성과 물리적 특성에 의해 기능성 식품과 의약품뿐만 아니라 주요 산업소재로 사용될 전망이다. 미국인을 위한 식사지침에 보면 각종 대사질환과 성인병을 예방하기 위해 1일 25g 이상의 콩 단백질 섭취를 권장하고 있다.

콩은 실로 우리 민족의 위대한 유산임에 틀림없다. 콩으로 만든 장(醬)은 우리 음식의 맛을 결정하는 기본 소재일 뿐만 아니라 간장을 먹는 동북아시아 문화권을 규정하는 인류문화사의 한 지표가 되는 것이다. 현대식 밥상 위에 거칠은 토기에서 부글부글 끓고 있는 된장, 청국장찌개를 올려놓고 먹는 우리의 유별난 식습관은 8000년이 넘는 우리의 독특한 음식문화를 보여주는 것이다.

영양적으로 콩은 한국인의 밥상에서 절대적 위치를 차지하고 있는 작물이다. 한국인은 쌀(밥)과 콩(된장찌개, 콩나물, 두부)만 있으면 영양학적으로 균형 잡힌 맛있는 한끼 식사를 할 수 있다. 우리의 콩음식은 한국 음식문화사의 기원과 형성 과정의 골격을 이루는 중심 소재이다.

영양인류학적 관점에서 보면 콩의 식용은 동이족이 문화적으로나 군사적으로 이 지역의 패권자로 군림하는데 결정적인 역할을 하였다. 실제로 중국 고대사에서 동이족은 한족(漢族)과 대등한 세력을 행사하였던 거대 민족으로 중국문명의 이른 시기에 주도적인 역할을 해왔음을 여러 문헌에서 발견할 수 있다. 『자원(字源)』에는 『후당서』를 인용하여 동이인(東夷人)은 중원지방에 먼저 살던 토착민으로 온후한 덕성과 도리를 분별하는 문화인으로 기술하고 있다.

일부 고대사 학자들은 은(殷)나라를 동이족이 건설한 초기 부족국가로 보고 있다. 『한서지리지』에는 동이를 천성이 유순하여 주변의 오랑캐들과는 다르다고 기재하고 있다. 타 민족을 오랑캐로 폄하하고 자만심이 강한 중국인들의 표현으로는 이례적인 것이다. 콩을 먹음으로 해서 동북아의 맹주가 되었던 우리 민족에게 콩은 여전히 쌀 다음으로 중요한 식량으로 반드시 자급해야하는 한국인의 기본 식량이다.

우리 민족에게 콩은 잊혀진 한국 고대사를 말없이 보여주는 증인이다. 그런 의미에서 영주시에 건립된 콩세계과학발물관은 콩 그 자체를 넘어서는 한민족의 역사 교과서이며 한국인의 세계 비전을 보여주는 곳이다.

식용 콩 자급을 위한 통일의 비전

식용 콩의 자급은 한국인의 식량안보를 위해 반드시 이루어야할 국가적 과제이다. 통일 이후 한반도 7천만 인구의 식용 콩 수요량은 62만 5,000톤으로 추산된다.[1] 북한은 산악지대가 많고 밭 면적이 넓어 콩 생산의 적지이다. 2009년 기준으로 남한의 콩 경지면적은 7만 5,000ha이며 북한은 9만ha이다. 콩의 수량성은 남한이 1.65톤/ha인 반면 북한은 1.16톤/ha로 남한의 70% 수준이다. 북한의 옥수수밭 50만ha 중 21만ha를 콩밭으로 전환하고 콩 수량성을 남한 수준으로 끌어올리면 남북한 총 37만 5,000ha에서 62만 5,000톤의 식용 콩을 생산할 수 있다.

1) 이철호, 문헌팔, 김세권, 박태균, 권익부, 한반도 통일과 식량안보, 도서출판 식안연 (2012)

통일 후 식용 콩 자급계획

구분	남한	북한	합계
식용 콩 수요량 (2010년, 천 톤)	438	187	625
식용 콩 생산량 (2010년, 천 톤)	139	154	293
콩 재배면적 (2010년, 천ha)	71	90	161
통일 후 재배면적 (천ha)	80	300*	380
통일 후 식용 콩 생산량 (천ha)	132	495**	627

*북한의 옥수수 재배면적 50만ha 중 21만ha를 콩밭으로 전환

**북한의 콩 수량성(1.16톤/ha)을 남한수준(1.65톤/ha)으로 향상